Dielectrics 2011

Journal of Physics: Conference Series Volume 310

Canterbury, United Kingdom
13-15 April 2011

Editors:

Alun Vaughan **Paul Lewin**

ISBN: 978-1-61839-130-8
ISSN: 1742-6588

Printed from e-media with permission by:

Curran Associates, Inc.
57 Morehouse Lane
Red Hook, NY 12571

Some format issues inherent in the e-media version may also appear in this print version.

Printed by Curran Associates, Inc. (2011)

For permission requests, please contact the Institute of Physics
at the address below.

Institute of Physics
Dirac House, Temple Back
Bristol BS1 6BE UK

Phone: 44 1 17 929 7481
Fax: 44 1 17 920 0979

techtracking@iop.org

Additional copies of this publication are available from:

Curran Associates, Inc.
57 Morehouse Lane
Red Hook, NY 12571 USA
Phone: 845-758-0400
Fax: 845-758-2634
Email: curran@proceedings.com
Web: www.proceedings.com

TABLE OF CONTENTS

PREFACE

In 2011, the biennial meeting of the Dielectrics Group of the IOP, *Dielectrics 2011*, was held for the first time in a number of years at the University of Kent at Canterbury. This conference represents the most recent in a long standing series that can trace its roots back to a two-day meeting that was held in the spring of 1968 at Gregynog Hall of the University of Wales. In the intervening 43 years, this series of meetings has addressed many topics, including dielectric relaxation, high field phenomena, biomaterials and even molecular electronics, and has been held at many different venues within the UK. However, in the early 1990s, a regular venue was established at the University of Kent at Canterbury and, it this respect, this year's conference can be considered as "Dielectrics coming home".

The format for the 2011 meeting followed that established at Dielectrics 2009, in breaking away from the concept of a strongly themed event that held sway during the mid 2000s. Rather, we again adopted a general, inclusive approach that was based upon four broad technical areas:

- Theme 1: Insulation/HV Materials
- Theme 2: Dielectric Spectroscopy
- Theme 3: Modelling Dielectric Response
- Theme 4: Functional Materials

The result was a highly successful conference that attracted more than 60 delegates from eight countries, giving the event a truly international flavour, and which included both regular and new attendees; it was particularly pleasing to see the number of early career researchers at the meeting. Consequently, the organizing committee would like to thank our colleagues at the IOP, the invited speakers, our sponsors and all the delegates for making the event such a success. Finally, we look forward to convening again in 2013, when we will be returning to The University of Reading.

Prof Alun Vaughan and Prof Paul Lewin, *Editors*

Broadband dielectric relaxation spectra of PVDF copolymer films

Bjoern Martin, Manfang Mai, and Herbert Kliem

Institute of Electrical Engineering Physics, Saarland University, Campus A5 1, D-66123 Saarbruecken, Germany

E-mail: b.martin@mx.uni-saarland.de

Abstract. The dielectric permittivity of polyvinylidenefluoride (PVDF) and its copolymer containing trifluoroethylene (TrFE) is investigated in the frequency domain with variation of the temperature. A pronounced relaxational behavior is observed in the films at high frequencies. In thick commercial PVDF films a Vogel-Tamman-Fulcher (VTF) dependence of the relaxation time versus the temperature is found yielding a thermally activated process. In thin Langmuir-Blodgett P(VDF-TrFE) films the relaxation times determined from the dielectric permittivity in the frequency range are not consistent with the relaxation times determined from the temperature range. The data extracted from the temperature domain also here show a VTF behavior. Nevertheless, the relaxation times calculated from the frequency domain cannot be fitted with the VTF formula. Thus, the time-temperature superposition principle is not valid in the thin films.

1. Introduction

Polyvinylidenefluoride (PVDF) and also its copolymers containing trifluoroethylene (TrFE) have gained much interest during the past years (see references in [1]). Due to the possibility to produce ultra-thin films down to 1 nm with a Langmuir-Blodgett deposition technique the material is a promising candidate for use in integrated electronic circuits. Since the polarization can be switched between two stable states with an external applied field it should be possible to employ P(VDF-TrFE) as nonvolatile ferrolectric access memory device. In former experiments it was demonstrated that also the thin films show ferroelectric properties like hysteresis loops with a remanent polarization of about 10 $\mu C/cm^2$ [2].

Small signal broadband dielectric spectroscopy measurements of PVDF and its copolymer in the frequency domain reveal a pronounced relaxational behavior at frequencies of about 10 MHz close to room temperature [1,3-6]. The temperature dependent behavior of the relaxation time is investigated here for different samples.

2. Samples preparation and experimental procedure

At first commercially available foils of pure PVDF with thicknesses of 9 μm and 22 μm are investigated. Aluminum electrodes are evaporated to both sides of the samples. With this sample geometry a simple measurement of the broadband dielectric relaxation spectra is possible.

Also, thin films of P(VDF-TrFE) copolymer are prepared by a horizontal Langmuir-Blodgett (LB) technique. This method allows the transfer of copolymer monolayers [7]. On the top of a glass substrate Al stripes with a width of 200 μm are evaporated (Fig. 1). Afterwards, the copolymer is transferred to the substrate. Al electrodes perpendicular to the first ones are evaporated to the copolymer film then. The crossing area of the Al stripes determines the active sample area.

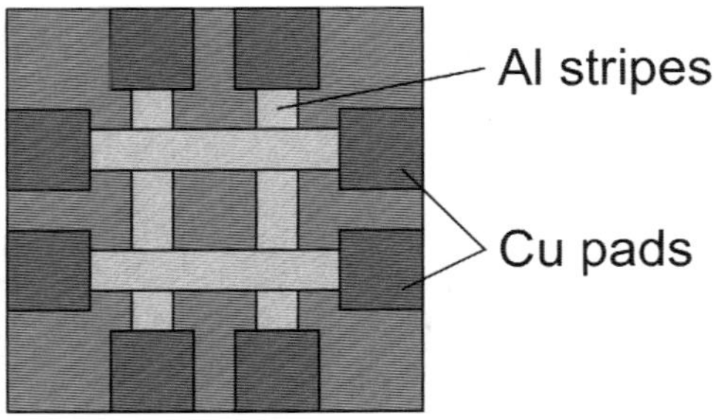

Figure 1. Sample structure. Thick Cu pads are evaporated on the Al stripes to allow a soldered contact.

The measurements in the frequency range are performed with an Agilent 4294A impedance analyzer for frequencies ranging from 40 Hz to 110 MHz and with an Agilent 4291A impedance analyzer from 1 MHz to 1.8 GHz. In the low frequency range below 1 kHz a self made analyzer system is used. All measurements are carried out in dry atmosphere in shielded measurement cells which can be cooled down to 200 K and heated up to 400 K. The thick commercial samples are placed in a coaxial line adjusted to the characteristic impedance of the measurement system which allows measurements up to 400 MHz. For the LB films onto the Al stripes Cu pads are evaporated (Fig. 1). The inner conductor of a coaxial line is directly soldered to the Cu pads. It should be noted that the use of contacting needles or silver paste yields measurement errors and unreproducible results when the temperature is varied.

During the heating and cooling cycle each 10 K a measurement of the complete frequency range is performed to find the data in the frequency range with a variation of the temperature, $\epsilon^*(\omega)$. In the same run the temperature and the dielectric permittivity at selected frequencies are measured each 30 s to get the data in the temperature range with variation of the frequency, i.e. $\epsilon^*(T)$. The relaxation time is identified by the maximum of the imaginary part of the dielectric permittivity either in the frequency or in the temperature range.

3. Experiments

3.1. Measurements performed with thick samples

Broadband dielectric spectroscopy measurements are performed using the commercial available PVDF foils. A pronounced relaxation is found in the high frequency range [6]. The relaxation time is shifted to shorter values with increasing temperature. The relaxation spectra is not Debye like but broadened especially at lower temperatures. An Arrhenius representation of the relaxation times for two different sample thicknesses (9 μm and 22 μm) is shown in figure 2. As can be seen the data for the two different sample thicknesses are nearly identical. The data determined from $\epsilon''(\omega)$ and from $\epsilon''(T)$ respectively are consistent. The data can be fitted with the well-known Vogel-Tamman-Fulcher (VTF) equation

$$\tau = \tau_0 \cdot exp\left(\frac{B}{T - T_0}\right)$$

with the VTF temperature $T_0 = 184$ K and the constants $B = 1170$ K and $\tau_0 = 2.6 \cdot 10^{-12}$ s. The time-temperature superposition principle is fulfilled for these samples.

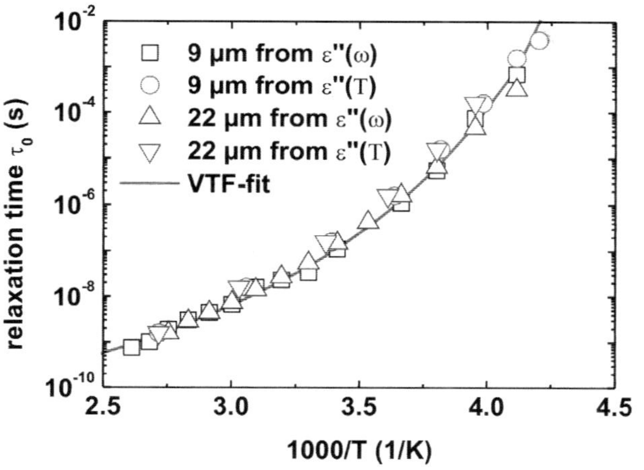

Figure 2. Arrhenius plot of the relaxation times for the 9 μm and 22 μm thick PVDF samples in dependence of the temperature calculated from $\epsilon''(\omega)$ and $\epsilon''(T)$. The relaxation times coincide and a fit with the VTF formula is possible.

3.2. Measurements performed with thin samples

In the next step three LB samples with different thicknesses (28 nm, 45 nm, and 69 nm) are investigated. The dielectric relaxation spectra in the frequency range with variation of the temperature during the heating process for the 28 nm sample are shown in figures 3(a) and 3(b). Also for these samples a dielectric relaxation is found in the high frequency range [1,3-5]. The relaxation time depends on the temperature and decreases with increasing temperature.

From the same set of data the dielectric permittivity in the temperature range is determined with variation of the frequency. The real part of the dielectric permittivity shows a typical ferroelectric behavior (Fig. 3(c)). During heating ϵ' increases up to the Curie temperature $T_{C1} \approx 380$ K in the ferroelectric phase. In the paraelectric region ϵ' decreases. If the sample is cooled afterwards the dielectric permittivity increases again down to a temperature $T_{C2} \approx 340$ K. Below T_{C2} the dielectric permittivity decreases with decreasing temperature. Thus, ϵ' shows a temperature hysteresis as expected for type I ferroelectrics. The hysteresis and the dielectric relaxation are also found in the imaginary part of the dielectric permittivity (Fig. 3(d)). All further data are extracted during the heating process. However, the high frequency dielectric relaxation time is not influenced by the direction of the temperature variation. This is a first indication that the dielectric relaxation is not directly caused by the ferroelectric system.

In figure 3(e) the relaxation times for different sample thicknesses calculated from the maximum of $\epsilon''(T)$ in the temperature range are presented. The values for all samples coincide and a fit of the data with the VTF formula ($\tau_0 = 2 \cdot 10^{-10}$ s, $T_0 = 195$ K, and $B = 750$ K) is possible. On the other hand the relaxation times observed from the maximum of $\epsilon''(\omega)$ in the frequency range for the 69 nm and the 28 nm sample show a curvature which cannot be fitted by VTF (Fig. 3(f)). Here, a fit with a simple Arrhenius formula seems to be possible if the temperatures below 240 K are neglected where the P(VDF-TrFE) undergoes the glass transition in thick samples [6]. Furthermore, the relaxation times presented in figure 3(e) differ from the values determined from $\epsilon''(T)$ shown in figure 3(f). Thus, the time-temperature superposition principle is not valid in the thin films.

Now, the question about the physical origin of the dielectric relaxation arises. In measurements performed on ferroelectric $BaTiO_3$ a comparable relaxational behavior is observed at frequencies of about 1 GHz at room temperature [8]. This behavior is attributed to a motion of domain walls. Because the relaxation in P(VDF-TrFE) is also present in the paraelectric phase a vibration of the domain walls can be excluded. Additionally, the relaxation time is independent of the polarization state, i.e. the unpolarized sample shows the same relaxation time as a completely polarized sample where no domains and thus no domain walls are present

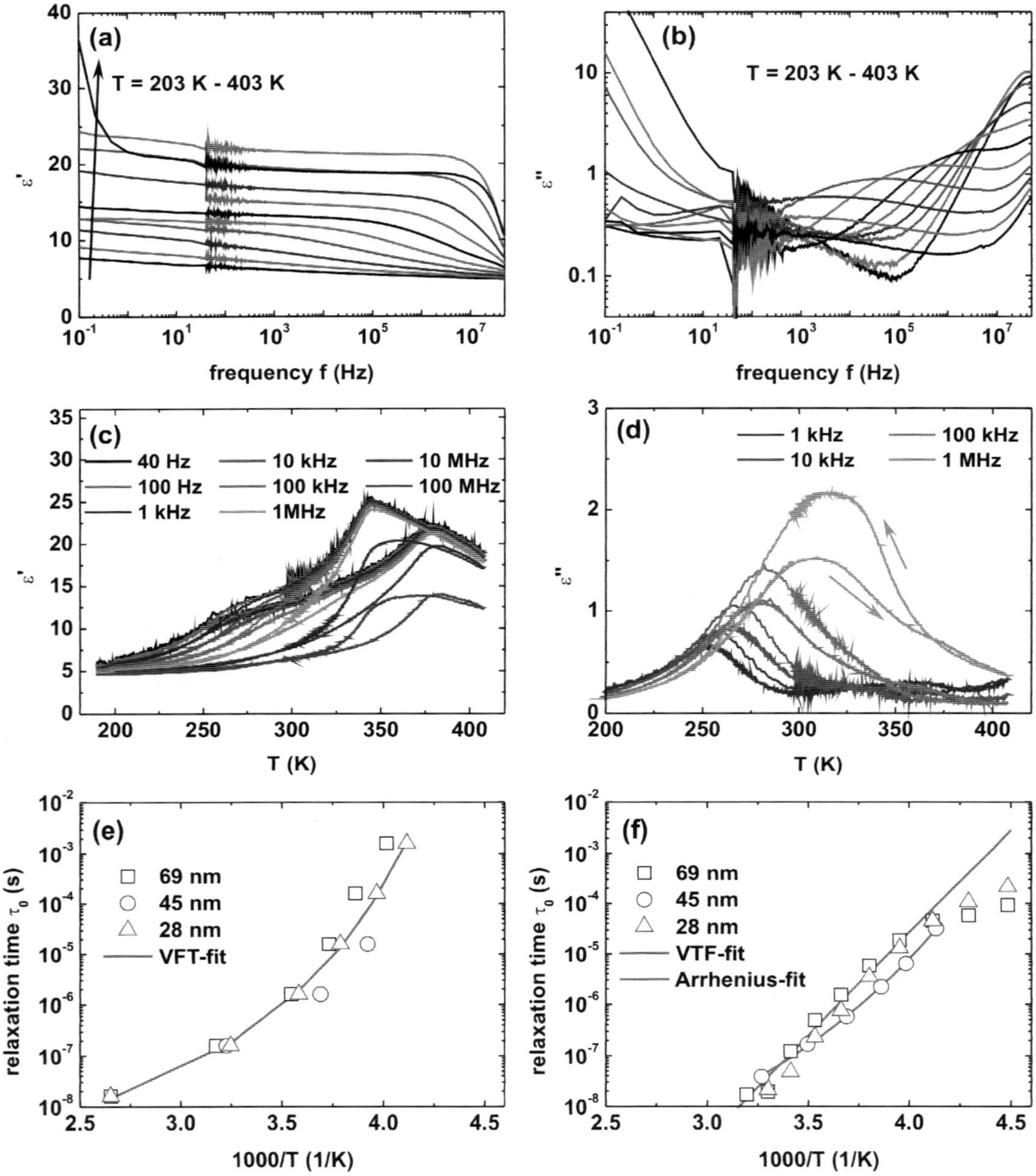

Figure 3. (a) Real and (b) imaginary part of the dielectric permittivity for the 28 nm thick LB sample in the frequency domain with variation of the temperature during the heating process. (c) Real and (d) imaginary part of the dielectric permittivity for the same data set in the temperature range with variation of the frequency. Next figures: Arrhenius plot of the relaxation times for three different LB samples calculated (e) from the maximum of $\epsilon''(T)$ in the temperature range and (f) from the maximum of $\epsilon''(\omega)$ in the frequency range. Different relaxation times are found and a fit with the VTF formula is not possible for the data calculated from $\epsilon''(\omega)$.

in the ideal case.

Thus, the physical origin of the relaxation and the explanation of the temperature dependence of the relaxation time in the LB films still remains an open question.

4. Conclusion

It is shown that a thermally activated relaxational behavior is found in thick PVDF as well as in thin LB P(VDF-TrFE) films. In the thick samples the relaxation times do not depend on the sample thickness and follow a VFT law in the temperature range. The thin samples show different behavior for the relaxation times versus the temperature determined from the maximum of $\epsilon''(\omega)$ compared to the data determined from the maximum of $\epsilon''(T)$. Thus, the time-temperature superposition principle is not valid in the thin films whereas it does hold for the thick films. Additionally, a domain wall motion can be excluded as the physical origin of the dielectric relaxation.

References

[1] Martin B, Kuehn M and Kliem H 2010 *J. Appl. Phys.* **108** 084109
[2] Kliem H and Tadros-Morgane R 2005 *J. Phys. D: Appl. Phys.* **38** 1860
[3] Jogad M S and Ducharme S 2002 *Curr. Sci.* **83** 472
[4] Verkhovskaya K A, Plakseev A A, Lotonov A M and Gavrilova N D 2009 *Ferroelectrics* **379** 144
[5] Martin B, Vizdrik G and Kliem H 2009 *J. Appl. Phys.* **105** 084114
[6] Solnyshkin A V and Kislova I L 2010 *Ferroelectrics* **398** 77
[7] Palto S, Blinov L, Bune A, Dubovik E, Fridkin V, Petukhova N, Verskovskaya K and Yudin S 1995 *Ferroelectr., Lett. Sect.* **19** 65
[8] Poplavko Y M, Tsykalov V G and Molchanov V I 1968 *Soviet Physics - Solid State* **10** 2708

System identification algorithms for the analysis of dielectric responses from broadband spectroscopies

S. Hadjiloucas[1], G.C. Walker[1], J.W. Bowen[1], and R.K.H. Galvão[2]

[1]Cybernetics, School of Systems Engineering, The University of Reading, RG6 6AY, UK

[2]Divisão de Engenharia Eletrônica, Instituto Tecnológico de Aeronáutica, São José dos Campos, SP, 12228-900 Brazil.

Email: s.hadjiloucas@reading.ac.uk

Abstract. We discuss the modelling of dielectric responses for an electromagnetically excited network of capacitors and resistors using a systems identification framework. Standard models that assume integral order dynamics are augmented to incorporate fractional order dynamics. This enables us to relate more faithfully the modelled responses to those reported in the Dielectrics literature.

1. Introduction

Over the past several years, our group has been building Fourier transform spectrometers for the far-infrared part of the spectrum and as a consequence, we have developed signal processing algorithms for de-noising and classification by decomposing the signals in the Fourier, windowed-Fourier, principal component and wavelet domains [1-3]. Furthermore, we have adopted a systems identification framework [4-7] for the direct analysis of signals in the time domain, obviating the commonly used ratioing procedure to evaluate the sample's insertion loss. Our modeling approach has been only valid, however, when the dispersive phenomena are governed by de-embeddable Lorentzian, Debye and Drude responses.

In addition, a significant body of literature has come to our attention, where the dielectric responses (frequency dependent permittivity and AC conductivity) of heterogeneous materials such as glasses, ceramics, amorphous semiconductors, heavily doped crystals, polymers, and composites can be modeled under the assumption that they behave as micro-structured electrical networks containing capacitive (insulating) and resistive (conducting) regions [8-18]. Often, the experimentally observed AC conductivity $\sigma(\omega)$ follows a Jonscher [19, 20] power law $\sigma(\omega)=\sigma(0)+A\omega^q$ where ω is the excitation angular frequency, $\sigma(0)$ is the DC conductivity, A is a proportionality constant and $0.5<q<0.7$. From this power law, as $\omega\to\infty$, the well-known irrational Cole-Davidson formula for complex permittivity is obtained:

$$\varepsilon^* \approx \frac{1}{\left(1+i\omega\tau\right)^{(1-q)}} \tag{1}$$

Since in the Laplace domain, the above expression is not of integral order, we propose the use of fractional order models [21] to describe such processes. The Riemann-Liouville definition of fractional integral and corresponding fractional order derivative is adopted to solve the corresponding fractional order differential equations of the underlying system dynamics:

$$ {}_0d_t^{-q} f(t) = \frac{1}{\Gamma(q)} \int_0^t (t-\tau)^{q-1} f(\tau) d\tau, \quad q \geq 0 \tag{2a} $$

$$ {}_0d_t^{p} f(t) = \frac{1}{\Gamma(1-p)} \int_0^t (t-\tau)^{-p} f(\tau) d\tau, \quad 0 < p < 1 \tag{2b} $$

where $\Gamma(q)$ denotes the Euler gamma function.

2. Signal processing schemes within a fractional order identification framework

The dynamics of an unknown medium may be modeled using an input-output state space representation and a subspace multiple input-multiple output (MIMO) model may be fitted [21]. The fractional state space representation for the system is given from:

$$ D^a \mathbf{x}(t) = A\mathbf{x}(t) + B\mathbf{u}(t) \tag{3a} $$

$$ \mathbf{y}(t) = C\mathbf{x}(t) + D\mathbf{u}(t) \tag{3b} $$

where $\mathbf{x} \in \mathbb{R}^n$ is the state vector, $\mathbf{u} \in \mathbb{R}^m$ is the input vector, $\mathbf{y} \in \mathbb{R}^p$ is the output vector and $A \in \mathbb{R}^{n \times n}$, $B \in \mathbb{R}^{n \times m}$, $C \in \mathbb{R}^{p \times n}$ and $D \in \mathbb{R}^{p \times m}$ are the state, input, output and feed-through system constant matrices to be determined. As discussed in [22], such fractional order system is stable if $0 < \alpha < 2$ and $|\arg(\lambda_k)| > \alpha\pi/2$ and $-\pi < \arg(\lambda_k) \leq \pi$ where λ_k corresponds to the kth eigenvalue of A. For time-domain simulations of a system, a recursive distribution of poles and zeros is obtained to approximate the frequency behavior of s^a over an interval $[\omega_A, \omega_B]$. Because the asymptotic behavior at the low and high limits of the above frequency interval can have a static error between the fractional order model and its approximation, it is common to minimize this using an integrator operating outside that interval.

Given a large number of inputs and outputs related to the unknown dielectric system, the goal of the subspace algorithms is to determine the order of the system, A, B, C, D (to a similarity transformation). For simplicity, we are only considering the deterministic case where there is no noise in the measured inputs or process (estimated state) although the proposed approach is also valid to the stochastic case where an explicit augmented model in innovation form can be considered. In that case, an additional step is required to obtain an estimate of covariance matrices of the noise sequences.

A linear order model is obtained after computing the successive α-order fractional derivatives of (3). After substitution, an extended linear model $\bar{\mathbf{y}}(t) = \Gamma_i^* \mathbf{x}(t) + \Delta_i^* \bar{\mathbf{u}}(t)$ is obtained with input and output variables:

$$ \bar{\mathbf{u}}(t) = \left[\mathbf{u}(t)^T \quad D^\alpha \mathbf{u}(t)^T \quad \ldots \quad D^{i\alpha} \mathbf{u}(t)^T \right]^T \tag{4a} $$

$$ \bar{\mathbf{y}}(t) = \left[\mathbf{y}(t)^T \quad D^\alpha \mathbf{y}(t)^T \quad \ldots \quad D^{i\alpha} \mathbf{y}(t)^T \right]^T \tag{4b} $$

where

$$ \Gamma_i^* = \left[C \quad CA \quad \cdots \quad CA^{i-1} \right]^T \quad \in \mathbb{R}^{ip \times n} \tag{5a} $$

$$ \Delta_i^* = \begin{bmatrix} D & 0 & \cdots & 0 \\ CB & D & \ddots & \vdots \\ \vdots & \ddots & \ddots & 0 \\ CA^{i-2}B & \cdots & CB & D \end{bmatrix} \quad \in \mathbb{R}^{ip \times im} \tag{5b} $$

correspond to the observability and controllability matrices.

Because a successive differentiation of the input-output data introduces additional noise from the high-pass filter action, a fractional order low-pass filter with corner frequency ω_f is introduced $F(s) = \left(1 + \tau s^\alpha\right)^{-1}$ where $\tau = \left(\omega_f\right)^{-\alpha}$ to provide numerical stability. This filter multiplies the Laplace transform of (3) and the result is inverse Laplace transformed to the time domain to obtain:

$$ \mathbf{x}(t) = A_F \left[\lambda \mathbf{x}(t) \right] + B_F \left[\lambda \mathbf{u}(t) \right] \tag{6a} $$

$$ \mathbf{y}(t) = C\mathbf{x}(t) + D\mathbf{u}(t) \tag{6b} $$

where $A_F = I + \tau A$, $B_F = \tau B$ and $\lambda = \Im^{-1}\left(F(s)\right)$ with $\Im$ denoting the Laplace operator and I the identity matrix. Using recursion on expression (6b), new filtered input and output vectors are constructed as well as observability and controllability matrices:

$$\mathbf{U}(t) = \left[\lambda^{i-1}\mathbf{u}(t) \quad \lambda^{i-2}\mathbf{u}(t) \quad \ldots \quad \mathbf{u}(t)\right]^T \tag{7a}$$

$$\mathbf{Y}(t) = \left[\lambda^{i-1}\mathbf{y}(t) \quad \lambda^{i-2}\mathbf{y}(t) \quad \ldots \quad \mathbf{y}(t)\right]^T \tag{7b}$$

$$\Gamma_i = \left[C \quad CA_F \quad \cdots \quad CA_F^{i-1}\right]^T \quad \in \mathbb{R}^{ip\times n} \tag{7c}$$

$$\Delta_i = \begin{bmatrix} D & 0 & \cdots & 0 \\ CB_F & D & \ddots & \vdots \\ \vdots & \ddots & \ddots & 0 \\ CA^{i-2}B_F & \cdots & CB_F & D \end{bmatrix} \quad \in \mathbb{R}^{ip\times im} \tag{7d}$$

The subsequent estimation of the A,B,C,D matrices follows the standard MOESP subspace identification routines discussed in the literature [23-27] by applying them to expressions 7a-7d. For brevity reasons these will not be further discussed here.

The fractional derivative order $\alpha \in (0,2)$ is considered unknown and is estimated by minimizing:

$$\hat{\alpha} = \underset{\alpha\in(0,2)}{\arg\min} \frac{1}{2}\left\|\hat{\mathbf{y}}_c(\alpha) - \mathbf{y}_c\right\|_2^2 \tag{8}$$

where $\mathbf{y}_c$ is the vector obtained by concatenating the system outputs and $\hat{\mathbf{y}}_c(\alpha)$ is the vector obtained by concatenating the outputs of the state-space representation estimated with the subspace model.

3. Modelling of dielectric responses

Figure 1 shows a generic dielectric network composed of resistors and capacitors connected in either series or parallel configurations.

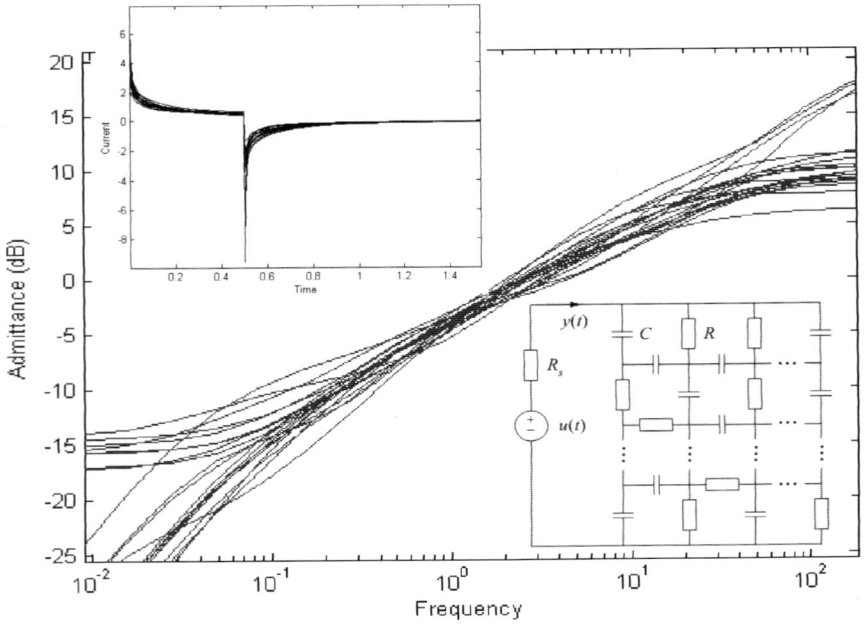

Figure 1. Simulated frequency response (admittance) of the dielectric network including the source resistance. Each line corresponds to one of the 20 random realizations assuming 5 hidden layers of nodes and 5 nodes per hidden layer. The insets shows the current pulse excitation and the random network of *R*, *C* components considered. The output resistance of the source is denoted by R_s.

The term "layer" will refer to a layer of nodes. The top and bottom layers are connected to the source. The intermediate layers are termed "hidden layers". The qualitative differences in the admittance responses can be ascribed to the position of the *R*, *C* components in the network. If there is any pure resistive path connecting the source with the ground, the low-frequency (DC) admittance will be different from zero. If there is no such path, the DC admittance will be zero (tending to minus infinity in dB). In the latter case, there is at least one capacitor blocking the DC current from the source to the ground. The presence/absence of pure resistive paths depends on the position of the components, which is chosen in a random manner. In the simulations performed, the elements have adopted values: $R_s = 0.1$, $R = 1$, $C = 0.5$ and different network realizations are considered.

The simulations always incorporate a resistor at the source output. Such a resistor can be used to model the output resistance of the source or the resistance of the interface between the source and the dielectric material. Moreover, the presence of such a resistor is important to eliminate inconsistencies in the model. For example, it makes no sense to connect the terminals of an ideal voltage source to a capacitor, as the capacitor voltage cannot undergo instantaneous changes. Following the work by Almond's group [10-14], the Matlab results presented here were further validated against those obtained by using the ORCAD 16.3 software to run SPICE simulations.

Figure 1 shows results obtained for 20 random networks (5 hidden layers of nodes, 5 nodes per hidden layer) in response to a rectangular pulse excitation $u(t)$ with duration 0.5 and unit amplitude. Figure 2 shows the time-domain current response of a particular network realization shown in the inset of figure 1 to a chirp excitation $u(t)$ with frequency ranging from 0 to 20 Hz and the corresponding N4SID models for integral order ranging from one to five.

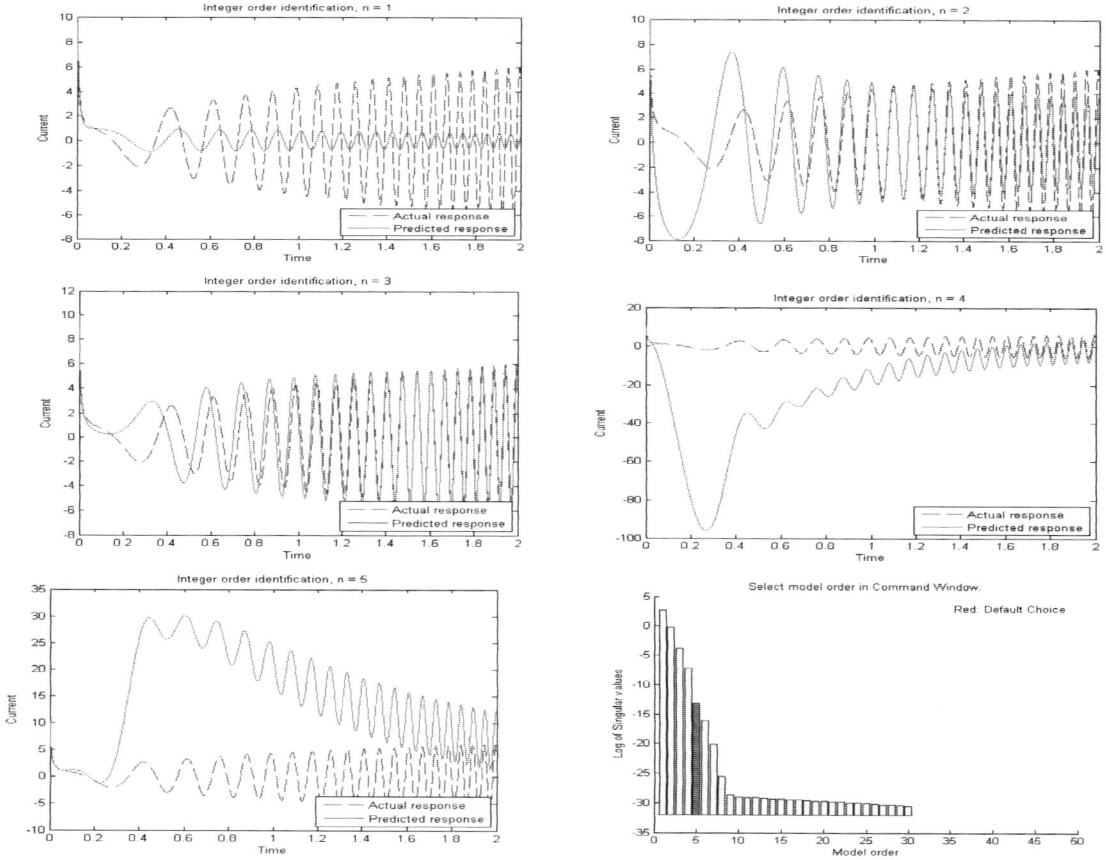

Figure 2. Current time domain response of a particular network to a chirp excitation $u(t)$ with frequency ranging from 0 to 20 Hz and N4SID models of integral orders n ranging from one to five.

These responses were obtained by stimulating the model dielectric with the same chirp input employed in the identification. In this case, the actual order of the input-output dynamics is 22. As can be seen, the difference between the actual and predicted responses decreases as the order is increased from one to three. However, the results become poor when the order is further increased, which may point to illconditioning problems in the identification process. Integral order subspace algorithms may not capture faithfully the dynamics of the dielectric sample under consideration.

Figure 3 presents the response of a fractional-order model identified by using customized simulations with a fractional order subspace algorithm. In this case, the identification was initialized by using the first-order N4SID model. As can be seen, the agreement between actual and predicted responses is a considerable improvement on the results of the integer-order models presented in Figure 2.

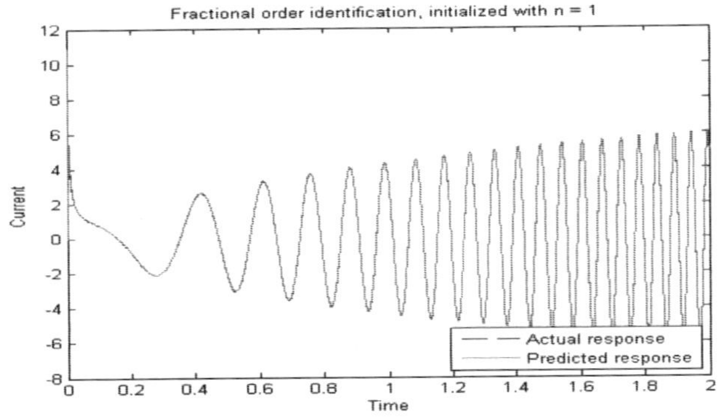

Figure 3. Time response of the identified fractional-order model.

A voltage-to-current transfer function $\left(0.098s^{0.88}+0.52s^{0.29}\right)\left(0.0096s^{0.91}+1\right)^{-1}$ for the identified fractional-order model was obtained for the particular realization used in the simulations (where the notation $s = j\omega$ implies that the transfer function is in the Laplace domain).

Figure 4 compares the frequency response of the best integer-order model (n = 3) and the fractional order model against the admittance of the actual circuit. As can be seen, the fractional-order model agrees with the expected response across a wider range of frequencies, which encompasses the spectral content of the chirp excitation employed in the identification procedure.

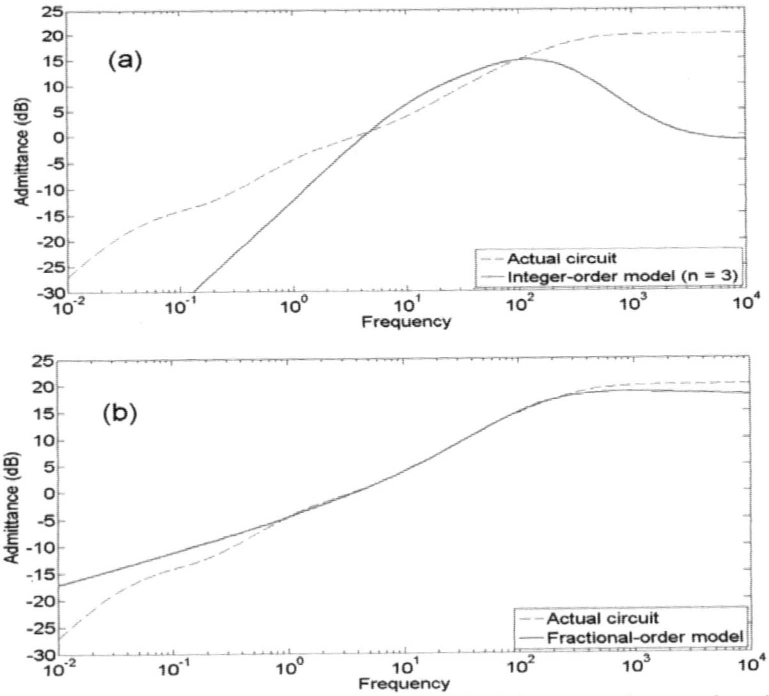

Figure 4. Frequency response of the identified models of (a) integer order n = 3 and (b) fractional order. Better matching between the actual circuit and the model response over 2 decades is obtained with the fractional order model.

4. Discussion

In the current framework, from an experimental perspective, we are assuming there are no additional artifacts from the propagation of the broadband excitation within the sample or from the process of gating the detector. It must be noted, however, that for the case of broadband excitation using a continuous wave Fourier transform spectrometer, where the sample is located in front of the detector after the two partial beams have recombined, there may be leakage of cross-polar components from the interaction of the sample with the beam [28] or pseudocoherence errors [29, 30] because different parts of the beam travel different paths through different regions of the sample (if this is of non-uniform thickness), interfering constructively or destructively with each other when they recombine. In addition, in the microwave and THz parts of the spectrum, in the case of continuous wave dispersive Fourier transform spectrometry, where the sample is located in one arm of the interferometer (here we are assuming a Michelson, Mach Zehnder or their polarization variance topologies) there could be some additional 'beam dilution' errors because the beam spreads diffractively at different rates along the axes of propagation in the reference and sample arms [31, 32]. This is further exacerbated in fast pulse spectroscopies where tight focusing of the IR beam is necessary unless an ASOPS detector interrogation scheme is utilized [8, 33]. Furthermore, in the case where a picosecond or femtosecond system has been used to electromagnetically excite the sample at the UV, optical, infrared, THz or microwave parts of the spectrum, there will be additional errors from the tight focusing of the beam to the sample or the detector and the beam may not be modeled under the assumption of plane wave incidence on the sample. Experimentally obtained dielectric responses often incorporate such artifacts and so the associated data-driven system identification models are expected to show deviations from theoretical responses. Interestingly, fractional order Fourier transforms can also be used to decompose the propagating EM beam modes inside the sample or in free space so the proposed formulation is also appropriate for the study of optical systems in general.

5. Conclusion

Jonscher's universal dielectric response of a random network of capacitors and resistors has been considered under a systems identification framework and a new approach using fractional order models is proposed. The scheme is particularly promising for modeling the responses of mixtures, where several species with different dielectric properties are present within the sample. It can also be used for de-noising purposes in a signal processing framework, as well as for sample classification purposes within a chemometric quality assurance context. Appropriate Matlab tools are currently under development to identify and validate fractional order models. The detailed mathematical formulation of fractional order identification tools and in particular the formulation of fractional order subspace models will be discussed at the conference.

References

[1] Galvão RKH, Hadjiloucas S, Bowen JW, and Coelho CJ, 2003 *Opt. Express*, **11**, 1462-1473.
[2] Hadjiloucas S, Galvão RKH and Bowen JW 2002 *J. Opt. Soc. Am. A* **19**, 2495-2509.
[3] Galvão, RKH, Hadjiloucas S, Zafiropoulos A, Walker GC, Bowen JW and Dudley R, 2007 *Opt. Lett.*, **32**, 3008-3010.
[4] Hadjiloucas S, Galvão RKH, Becerra VM, Bowen JW, Martini R, Brucherseifer M, Pellemans HPM, Haring Bolívar P, Kurz H, Chamberlain JM 2004 *IEEE Trans. Microw. Theory*, **52**, 2409-2419.
[5] Fischer BM, Yin X, Ng BW-H, Abbott D, Paiva HM, Galvão RKH, Hadjiloucas S, Walker GC, and Bowen JW, 2008 'Subspace and Wavelet-Packet Algorithms for de-noising and classifying broadband THz transients,' *Joint 33rd Int.Conf. Infrared and Millimetre Waves and 16th Int.Conf. Terahertz Electronics,* Caltech, USA.
[6] Yin X, Fischer BM, Ng BW-H, Abbott D, Paiva HM, Galvão RKH, Hadjiloucas S, Walker GC and Bowen JW 2007 In *Microelectronics: Design, Technology, and Packaging III,* (Eds. Alex J. Hariz, Vijay K. Varadan), Proc. SPIE **6798**, pp. 679814.
[7] Hadjiloucas S, Walker GC, Bowen JW, Becerra VM, Zafiropoulos A and Galvão RKH 2009 *J. Phys. Conf. Ser.*, **183** 012003
[8] Nowick AS, Vaysleyb AV and Kuskovsky I 1998 *Phys. Rev. B*, **58**, pp. 8398-8406.
[9] Panteny S, Stevens R and Bowen CR 2005 *Ferroelectrics*, **319**, pp. 199-208.
[10] Almond DP, Vainas B and Uvarov NF 1998 *Solid State Ionics*, **111**, pp. 253-261.
[11] Vainas B, Almond DP, Luo J and Stevens R 1999 *Solid State Ionics*, **126**, pp. 65-80.
[12] Almond DP, Bowen CR and Rees DAS 2006 *J. Phy. D. Appl. Phys.,* **39**, pp.1295-1304.

[13] Bouamrane R and Almond DP 2003 *J. Phys. Condens. Matt.*, **15**, pp. 4089-4100.
[14] McCullen NJ, Almond DP, Budd CJ and Hunt GW 2009 *J. Phys. D Appl. Phys.*, **42**, 064001.
[15] Dyre JC 1991 *J. Non-Crystal. Solids*, **135**, pp. 219-226.
[16] Wieczorek W, Zalewska A, Siekierski M and Przyluski J 1996 *Solid State Ionics*, **86-88**, pp. 357-362.
[17] Nowick AS and Lim BS 2001 *Phys. Rev. B*, **63**, 184115.
[18] Sidebottom DL 1999 *Phys. Rev. Lett.*, **83**, pp. 983-986.
[19] Jonscher AK 1977 'The universal dielectric response' *Nature* (London) 267, pp. 673-679.
[20] A.K. Jonscher AK *Dielectric Relaxation in Solids* (Chelsea Dielectrics Press, London, 1983).
[21] Hartley TT, Lorenzo CF, 2003 *Signal Process.*, **83**, pp. 2287-2300.
[22] Matignon D 1998 Stability properties for generalized fractinal differential systems, In *ESAIM: Proc. Fractional Differential Systems: Models, Methods and Applications*, **5**, 145-158.
[23] van Overschee P and De Moor B 1994 *Automatica* **30** 75–93
[24] Katayama T (2005), *Subspace methods for system identification*, Springer.
[25] Favoreel W, De Moor B, Van Overschee P 2000 *J. Process Contr.*, **10**, pp. 149-155.
[26] Viberg M, 1995, *Automatica*, **31**, 1835-1851.
[27] Galvão RKH, Hadjiloucas S, Becerra VM and Bowen JW 2005 *Meas. Sci. Technol.*, **16**, 1037-1053.
[28] Bowen JW, Walker GC and Hadjiloucas S. 2007 *Joint 32nd Int.Conf. Infrared and Millimetre Waves and 15th Int. Conf. Terahertz Electronics*, Cardiff, UK pp.208-209 IEEE Cat. number 07EX1863.
[29] Hadjiloucas S, Karatzas LS and Bowen JW 1999 *IEEE Trans. Microw. Theory*, **47**, 142-149.
[30] Walker 2003 GC '*Modelling the Propagation of Terahertz Radiation in Biological Tissue*' PhD Thesis, Centre of Medical Imaging Research, University of Leeds.
[31] Bowen JW, Walker GC, Hadjiloucas S and Berry EA 2004 'The consequences of diffractively spreading beams in ultrafast THz spectroscopy,' *Joint 29th International Conference on Infrared and Millimeter Waves and 12th International Conference on Terahertz Electronics*, Karlsruhe, Germany pp. 551-552, IEEE Catalog Number 04EX857.
[32] Walker GC, Bowen JW, Zafiropoulos A, Hadlington T, Hadjiloucas S, Chamberlain JM 2007 *Joint 32nd Int.Conf. Infrared and Millimetre Waves and 15th Int. Conf. Terahertz Electronics*, Cardiff, UK pp. 512-513. IEEE Cat. number 07EX1863.
[33] Bartels A, Cerna R, Kistner C, Thoma A, Hudert F, Janke C, and Dekorsy T 2007 *Rev. Sci. Instrum.* **78**, 035107.

Microscopic resolution broadband dielectric spectroscopy

S. Mukherjee, P. Watson, R. J. Prance

Centre for Physical Electronics and Quantum Technology, School of Engineering & Design, University of Sussex, Brighton, BN1 9QT, United Kingdom

r.j.prance@sussex.ac.uk

Abstract. Results are presented for a non-contact measurement system capable of micron level spatial resolution. It utilises the novel electric potential sensor (EPS) technology, invented at Sussex, to image the electric field above a simple composite dielectric material. EP sensors may be regarded as analogous to a magnetometer and require no adjustments or offsets during either setup or use. The sample consists of a standard glass/epoxy FR4 circuit board, with linear defects machined into the surface by a PCB milling machine. The sample is excited with an a.c. signal over a range of frequencies from 10 kHz to 10 MHz, from the reverse side, by placing it on a conducting sheet connected to the source. The single sensor is raster scanned over the surface at a constant working distance, consistent with the spatial resolution, in order to build up an image of the electric field, with respect to the reference potential. The results demonstrate that both the surface defects and the internal dielectric variations within the composite may be imaged in this way, with good contrast being observed between the glass mat and the epoxy resin.

1. Introduction

Electric potential sensors (EPS) are ultra-high input impedance devices which may be configured to measure spatial potential, electric field or surface charge with minimal loading to the electric field [1]. The principle of operation has already been described in detail in the literature by the authors [2]. The sensors have previously been applied to imaging applications and to the characterization of dielectric materials [3,4], but were only capable of operation at a single low audio frequency. The purpose of this paper is to present preliminary results for a broadband version of the sensor capable of operating from <10 kHz to >100 MHz with microscopic spatial resolution. Samples of FR4 printed circuit board have been chosen initially, because it is a well characterized composite dielectric material of considerable interest to the electronics community. Preliminary results are presented in the form of images of the spatial potential, close to the surface of the sample, over a range of excitation frequencies from 10 kHz to 11 MHz. From the microscopy point of view, small linear defects of varying widths have been milled on to the surface of the sample, the minimum width being ~200 μm, and the sense electrode being 50 μm in diameter.

2. Method

The sample of FR4 chosen for these measurements was a single sided copper clad board with two linear defects machined into the top surface of the dielectric using a PCB milling machine. The experimental setup is shown schematically in figure 1. The a.c. excitation signal is connected to the lower copper surface of the sample and the sensor electrode is positioned above the sample. The input impedance of the sensor is represented by the C_{in} and parallel resistance in the figure. Typically the effective input capacitance is of the order of femtofarads in parallel with terraohms. This extreme input impedance is of importance, because when using a microscopic sense electrode to couple to the

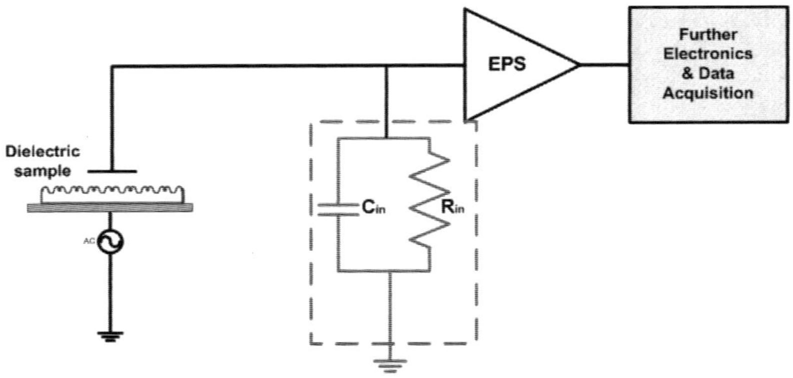

Figure 1. Schematic of electric potential sensor and detail of measurement.

sample the coupling capacitance is usually in the femtofarad range. If the input capacitance of the sensor is considerably larger than this it will act as a capacitive potential divider and attenuate the signal considerably. In addition if the input resistance is insufficient, this forms a high pass filter and will also attenuate the signal. The frequency response for the sensor is shown in figure 2 through a number of well defined coupling capacitors Cc, the smallest being 53 fF.

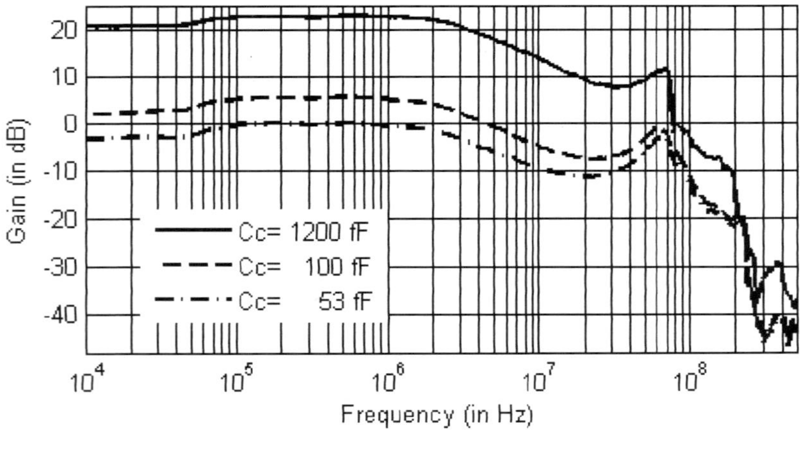

Figure 2. Frequency response of electric potential sensor as a function of coupling capacitance.

The sensor is raster scanned over the surface of the sample with a step size of 63 μm, at constant height comparable with the diameter as the sense electrode, in this case 50 μm. The spatial resolution of the image is comparable with these dimensions. Data is collected as a function of position to build up an image of the potential, using the same LabVIEW virtual instrument which controls the raster scanning. A simple algorithm is used to subtract background curvature from the data sets presented.

3. Results

Figure 3 shows a 3-D image of the potential ~50 μm above the surface of the sample at an excitation frequency of 10 kHz. Two things are immediately apparent from this result, first that the three machined linear defects are clearly visible as black lines and second that the internal structure of the material may be seen. This is manifest as the contrast seen between the resin (ε_r ~3.4) and the glass mat (ε_r ~6.0). The average dielectric constant for this composite material depends on a number of factors including the ratio of glass to resin and the frequency of the measurement. The average value has been measured and shown to decrease with increasing frequency [5], from $\varepsilon_r = 4.65$ at 10 kHz, to $\varepsilon_r = 4.25$ at 10 MHz. The defects milled on the sample surface in this case have a minimum width of ~200 μm, which as seen in the result, are clearly detectable.

Figure 3. 3-D rendering of surface potential with an excitation signal of 100 kHz. Two linear surface defects may be seen in addition to the internal structure of the material.

Figure 4a shows results obtained at a higher excitation frequency of 1 MHz where we observe a lower contrast between the glass mat and the resin than was seen at 100 kHz. The data is represented as 2-D images in this case with raw data at the top of the figure and processed data at the bottom. Figure 4b illustrates the image processing which has been applied to this data, a single line of raw data across the centre of the sample is shown at the top and the same line signal processed at the bottom. A simple fitting algorithm has been used to subtract the background curvature from the data set. This effect is caused by the surface curvature of the sample not being sufficiently well controlled at the 50 μm level. Variations in the sample to sensor spacing change the coupling capacitance and hence the effective gain, as may be clearly seen from figure 2.

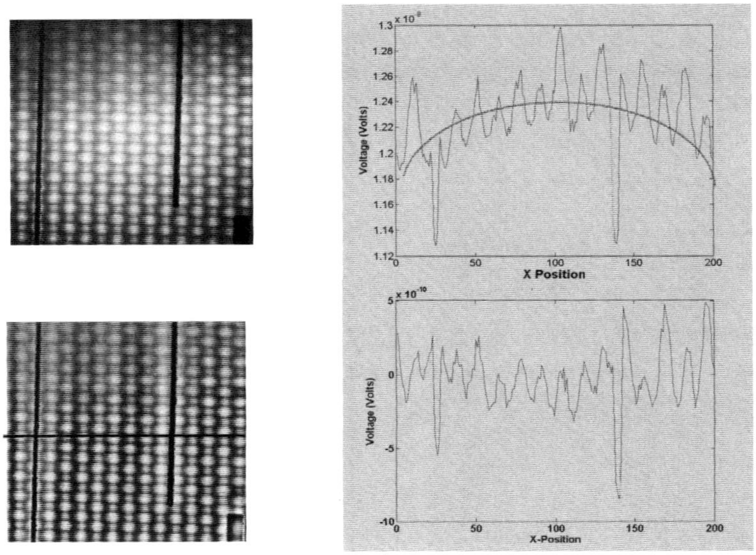

(a) (b)

Figure 4. (a) 2-D plots of potential at 1 MHz, (b) line plots of potential showing background curvature and detail of glass mat. Raw data (top) and data with background subtracted (bottom).

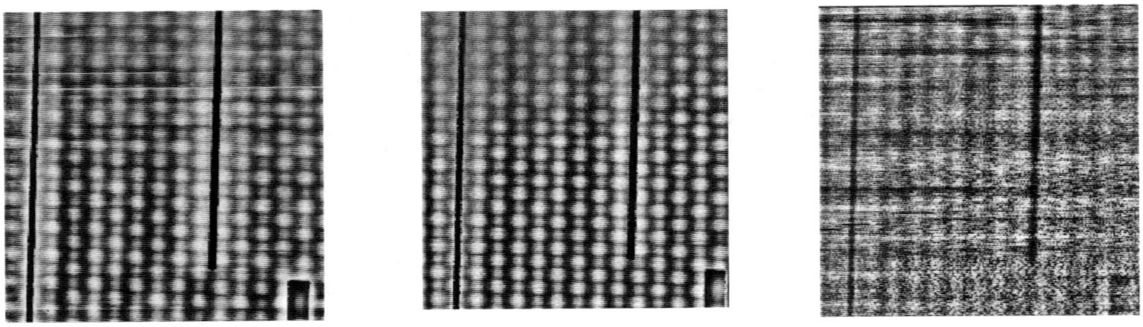

(a) (b) (c)

Figure 5. 2-D plots of surface potential at frequencies of: (a) 10 kHz, (b) 1 MHz and (c) 11 MHz.

In figure 5 data is presented as 2-D plots for three excitation frequencies, of 10 kHz, 1 MHz and 11 MHz. In all three cases the two machined linear surface defects are clearly visible, but there is a marked difference in the detail of the highest frequency image where the signal obtained from the sensor is relatively more attenuated resulting in poor dielectric contrast. However, the two machined defects and some internal structure are still discernable. This has resulted due to two reasons. It is worth noting that it has been reported that the loss factor for FR4 is a maximum in the 1-10 MHz region [5].

4. Conclusions

Preliminary results have been described for an imaging system based on the electric potential sensor and capable of producing images of the potential close to the surface of a dielectric material. The sample was excited via a copper backplane using an a.c. source. These images contain information about both the surface features and also about the local variations in dielectric properties in the material. A sample of FR4 circuit board was chosen for demonstration purposes as an example of a well characterised composite dielectric material. Results were obtained over a range of frequencies from 10 kHz to 11 MHz, with a spatial resolution of the order of 50 μm. In other applications we have already demonstrated the capability of achieving 1 μm resolution which also sets the limit for detection resolution of surface defects in dielectric materials [4], microscopic surface charge imaging [6] and the use of arrays of sensors to speed up the acquisition of images [7].Clearly, the detection resolution is essentially proportional to the spatial resolution and scales down with it making it possible to detect the sample defects of widths comparable to the dimension of the sense electrode.

Acknowledgements

The authors would like to thank the Engineering and Physical Sciences Research Council for supporting this work under Grant No. EP/E042864/1.

References

[1] Aydin A, Stiffell P B, Prance R J and Prance H, 2010 *Meas. Sci. & Technol.* **21** 125901

[2] Prance R J, Debray A, Clark T D, Prance H, Nock M, Harland C J and Clippingdale AJ 2000 *Meas. Sci. & Technol.* **11** 291-297

[3] Beardsmore-Rust S T, Watson P, Prance R J, Harland C J and Prance H 2009 *Meas. Sci. & Technol.* **20** 095711

[4] Clippingdale A J, Prance R J, Clark T D and Brouers F 1994 *J Phys* **D 27** 2426-2430

[5] Mumby S J and Jih Yuan 1989 *Journal of Electronic Materials*, **18**(2) 287-292

[6] Watson P, Prance R J and Beardsmore-Rust S T 2011 accepted for publication in *Forensic Science International*

[7] Gebrial W, Prance R J, Harland C J and Clark T D 2006 *Rev Sci Instrum* **77**(6) DOI: 10.1063/1.2213219

Integrated microwave resonant device for dielectric analysis of microfluidic systems

D J Rowe [1,2], **A Porch** [1], **D A Barrow** [1] **and C J Allender** [2]

[1] School of Engineering, Cardiff University, Cardiff CF24 3AA, UK
[2] Welsh School of Pharmacy, Cardiff University, Cardiff CF24 3NB, UK

Email: rowedj@cf.ac.uk

Abstract. Herein we present a device for performing non-contact dielectric spectroscopy upon liquids in a microfluidic environment. The device is comprised of a compression-sealed polytetrafluoroethylene (PTFE) chip with an embedded coaxial resonator, which is overmoded for dielectric measurements at six discrete frequencies between 1 and 8 GHz. A novel capacitive coupling structure allows transmission measurements to be taken from one end of the resonator, and an optimised microchannel design maximises sensitivity and repeatability. The use of a PTFE substrate and a non-contact measurement gives excellent chemical and biological compatibility. A simple 'fingerprint' method for identifying solvents is demonstrated, whereby a sample is characterised by air-referenced changes in complex frequency. Complex permittivity values are also obtained via a perturbation theory-based inversion. A combination of experimental and simulated results is used to characterise the device behaviour, limits of operation and measurement uncertainty. The high stability of temporal measurements, coupled with the robustness of the design, make this device ideal for analytical chemistry and industrial process control.

1. Introduction

Dielectric metrology has applications as an alternative and complimentary spectroscopic technique because of key advantages over other, more ubiquitous methods, including the potential for non-destructive, non-invasive and label-free measurements, which are not hampered by the presence of solid material within a liquid phase. This work describes the development of a device for liquid-phase dielectric measurements in a platform that enables *in situ* characterisation of chemical, biological and pharmaceutical processes. It is novel in three main respects: the resonator coupling structure, the use of an overmoded resonator to give multi-frequency information, and the microfluidic interface.

An integrated device – one comprised of a sensing technology and appropriate sample interface – is necessary for low volume, high throughput analytical procedures. Filling a waveguide with a solvent [1] or dipping a reflectance probe [2] into a solution are sufficient for several samples but not for the applications of modern chemistry and biology. Instead, the inclusion of microfluidic techniques enables the realisation of a more practical measurement system for such applications. Laminar flow can be exploited by appropriate micron-scale design, enabling the control of various parameters to a great degree of accuracy: concentration, temperature, mixing, etc. [3,4], whilst using extremely small volumes. Such a system is attractive because of its potential for low-cost, disposable, point-of-care devices, with the obvious safety benefits of a sealed device.

It is advantageous to focus dielectric measurements on the microwave region for solvent-based systems, because it is an interesting range where many dipolar relaxations can be observed [5-8], and where the effects of dissolved species on these relaxations can be seen. A resonant structure was chosen for greater accuracy and less uncertainty when compared with equivalent broadband techniques [9]. They are more limited in scope by being fundamentally restricted to measuring discrete frequencies rather than continuous spectra, but the use of an overmoded resonator counteracts this problem. A resonant structure is also less dependent on the Vector Network Analyser (VNA) calibration, resulting in reduced systematic error. Using a coaxial resonator is beneficial due to ease of integration: the resonator can be perturbed by being positioned near to a sample, but not necessarily in contact with it. There are also a large number of well-defined TEM modes before hybrid and parasitic modes appear in the spectrum. Semi-rigid coaxial cable was selected as it is cheap, easy to machine, a standard size, and sufficiently robust for industrial applications. Although a coaxial resonator will have a lower quality factor (Q) than cavity or dielectric resonators (where Q is the quotient of resonant frequency and half-power bandwidth), this disadvantage is outweighed by the advantages. Transmission rather than reflectance measurements were used for improved dynamic range.

2. Design

2.1. Perturbation Theory

The reader is referred to [10] for a full formulation of perturbation theory for microwave resonators, but a cursory description is provided here for sake of completeness. It should be noted that while the approach of [10] is formulated with respect to a cavity resonator, it is equally applicable for evanescent field perturbation at the end of a coaxial resonator.

A material can interact with the electromagnetic fields of the resonator according its complex values of permittivity and permeability, which interact with the electric and magnetic fields respectively. Due to the difficultly of decoupling the effects of these two parameters, this device is intended for measurements of non-magnetic (i.e. zero permeability) materials, and as such magnetic field interactions will not be considered.

The stored energy of the empty and perturbed fields are related to their volume integrals, and can also be related to a normalised change in complex frequency in a derivation beginning with Maxwell's equations [10], where complex frequency $\tilde{f}$, as defined in equation (1):

$$\tilde{f} = f_0 - i\frac{1}{2}f_B \quad (1)$$

where f_0 is resonant frequency and f_B is half-power bandwidth. If complex permittivity is isotropic and unchanging, it can be calculated from measurements of centre frequency and bandwidth assuming some knowledge of the stored energy. In practice, this can be achieved by assuming the resonator fields are changed by a negligible amount upon the introduction of a sample material, so their volume integrals can be assumed to be constant. The expression then approximates to show a change complex permittivity is linearly proportional to the resulting change in complex frequency, so unknown materials can be characterised according to a measurement of a known calibration material. However, the unchanging field assumption does not hold for most reasonable values of permittivity, particularly for the relatively lossy solvents of interest to the applications described in the introduction, so measurements must be calibrating with a number of solvents as described in section 3.

2.2. Overmoded resonator

The open-ended half-wavelength resonator was formed of open-circuited 50 Ω RG401 semi-rigid coaxial cable, which was cut to 8 cm to give a fundamental resonance of ~ 1.25 GHz. This resonance was chosen to give six modes in the frequency region of interest. It has copper conductors (6.35 mm outer diameter) and a PTFE spacer (inner diameter 1.6 mm, outer diameter 5.4 mm). These dimensions were chosen as a compromise between a suitably small aperture size for microfluidic channels, against resonator losses, which increase with frequency.

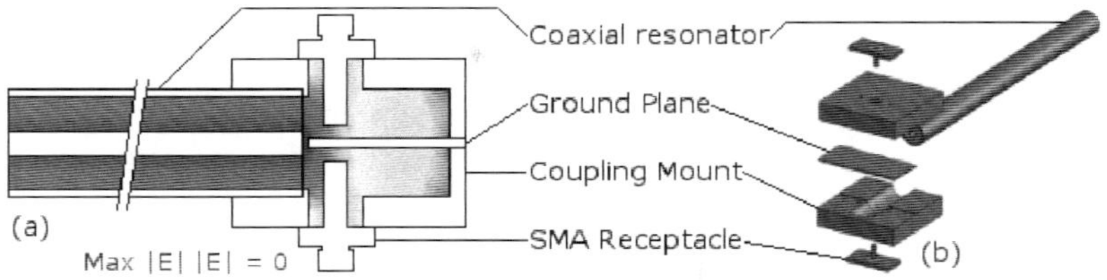

Figure 1. Coupling structure: (a) cross-sectional view of coupling structure with electric field magnitude colour map, |E| (Vm^{-1}), and (b) 3D exploded view.

2.3. Coupling structure
Transmission measurements require two ports, and are most easily achieved by coupling with parallel coaxial feedlines located a small distance from either end of the resonator. However, this structure is difficult to perturb with a sample – for example, it is possible to pass a capillary between the resonator and a feedline, but the tolerances on the inner and outer diameters of capillaries generally yield a +/- 10% uncertainty. Invasive couplings are also possible, whereby receptacles protrude orthogonally into the resonator, but these will not be able to couple to certain higher-order modes which have nulls adjacent to the connectors.

The cross-sectional and 3D views of figures 1 (a) and (b) show how transmission measurements can be taken from one end of the probe without having to couple at suboptimal points along the length of the resonator. The ground plane is necessary to minimise crosstalk between the connectors. This structure was designed to be weakly coupled, which means coupling has a minimal effect on the resonator fields, so it does not have to be accounted for during the inversion process. This design was simulated in COMSOL Multiphysics, and the relative dimensions and positions of the ground plane, receptacles and resonator were optimised to maintain weak coupling and maximise dynamic range.

2.4. Microfluidic interface
A 400 μm square channel, widening above the probe aperture, was micromilled in a PTFE chip. A recess in the back of the chip housed the resonator 100 μm below the channel in order to improve chemical and biological compatibility; to prevent fluid ingress into the resonator and to reduce corrosion of the copper conductors. Holes were drilled through the chip at either end of the channel to accept capillaries for fluidic inlet and outlet connections. These were fixed in place by standard fittings that screwed into the manifold. The use of a PTFE microfluidic setup meant that the system could be cleaned by flushing with acetone or other solvent, removing the need for disassembly between measurements and thus improving measurement reproducibility. A gasket layer of 100 μm PTFE film was placed between the machined chip and another, plain PTFE chip, and compression sealed in a brass manifold. The device can be added directly into existing flow systems because of the use of a PTFE substrate and standard fittings.

3. Experimental Methodology
The transmission coefficient was measured for each resonance with an Agilent E5071B VNA, and exported to LabVIEW, where a Lorentzian curve was fitted to the data in order to determine the centre frequency and bandwidth in real-time. A segmented frequency sweep was used to study all modes simultaneously. The frequency range of the VNA meant six modes could be measured with each sweep taking roughly 15 milliseconds. For each sample, air was pumped through the system and a reference measurement taken. Immediately following this, the sample solvent (HPLC-grade) was pumped through the device and a measurement again taken. This process was repeated 8 times for each solvent. The environment was maintained at constant temperature by passing a 1 m length of inlet capillary through a 20 °C water bath.

A 3D model of the integrated device was created in COMSOL, and simulations were used to calculate the centre frequency and bandwidth of each resonance within the frequency region of interest. A series of simulations were performed where resonant frequency and bandwidth were calculated as the real and imaginary components of sample permittivity were swept from $1 \rightarrow 30$ and $i0 \rightarrow i15$. Two important observations arose from investigating the relationships between these parameters. Firstly, each component of complex frequency could be seen to depend on both real and imaginary permittivity. This fact is ignored by the majority of groups that employ perturbation theory, which simply equate real and imaginary components. Secondly, the relationships between complex frequency and complex permittivity were non-linear. This is also commonly ignored, as the relationships between permittivity and frequency are normally treated as linear proportionalities, as described in section 2.1. An analytic inversion based upon perturbation theory was therefore deemed unsuitable as it would not be able to account for these behaviors. Instead, measurements of several known solvents over a range of real and imaginary permittivities were used to interpolate an inversion function for each mode from the associated changes in complex frequency. This was verified by comparing inverted measurements with previously obtained dielectric dispersion data, together with simulations of the same solvents.

4. Results and Discussion

The results were processed in two ways. The first takes the changes in the real and imaginary components of frequency (figure 2), where the change in complex frequency $\Delta\tilde{f}$ is defined in equation (2),

$$\Delta\tilde{f} = \Delta f_0 - i\frac{1}{2}\Delta f_{\mathrm{B}} = g(\varepsilon_1 - i\varepsilon_2), \qquad (2)$$

where ε_1 is permittivity, ε_2 is dielectric loss and g is a non-linear function of complex permittivity $\varepsilon_1 - i\varepsilon_2$. This method of post-processing could be used for real-time monitoring of a liquid sample as it does not require the interpolation of an inversion function. The variance of the data was ten parts in a million, so error bars are not appropriate. The error bars do not overlap in any instance, so there is no ambiguity in distinguishing solvents. This system also lends itself to multi-dimensional classification.

Permittivity spectra of the measured samples are shown in figure 3, where interpolated measurements of chloroform, ethanol, methanol, acetone and acetonitrile, were used to calculate the complex permittivity of toluene. The grey lines represent the dielectric relaxations to which the measured changes in complex frequency were fitted. Relaxation parameters were taken from [5-8].

Table 1 shows the following sensor parameters: limit of detection, geometric uncertainty and maximum standard error. The standard error was used to calculate the limit of detection; that is, the minimum change in complex frequency that could be discerned above measurement noise. The uncertainty was taken as the maximum of a series of correlated simulations modelling fabrication

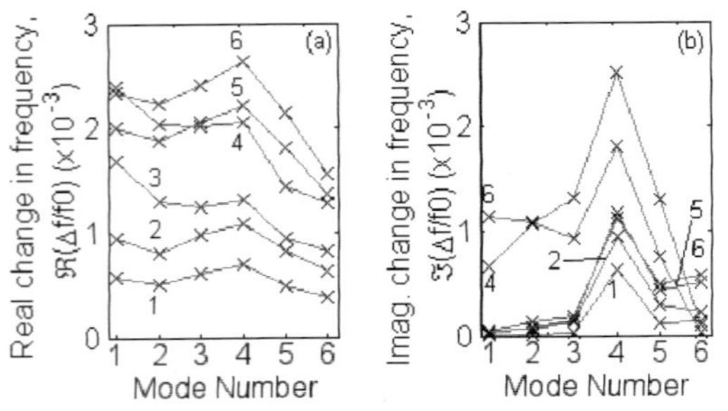

Figure 2. Changes in (a) real and (b) imaginary frequency for each mode of the resonant device, for (1) toluene, (2) chloroform, (3) ethanol, (4) methanol, (5) acetone and (6) acetonitrile.

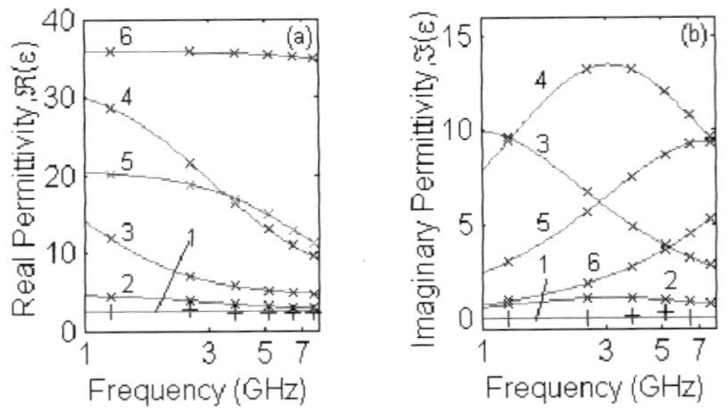

Figure 3. (a) real and (b) imaginary permittivity, for solvents as Figure 2. Solid lines represent Debye or Cole-Cole model of the corresponding solvent; 'x' represents fitted and '+' represents inverted measurements.

Table 1. Sensor Characteristics

Standard error	Limit of detection, ε_{DR}	Uncertainty
± 7 kHz	1.00011	± 1.7 %

tolerances, which can be assumed to be the dominant source of uncertainty.

Given the large range of permittivities and the small number of solvents, it is not appropriate to define a resolution with the available data. The device is suitable for compositional measurements (for industrial, chemical and pharmaceutical purposes) as it has been exercised over a large range of permittivity, yet still demonstrates a low error.

5. Conclusion

A new method of interfacing a microwave measurement technique with a microfluidic circuit has been demonstrated. Three novel aspects (the coupling structure, the overmoded resonator and the microfluidic interface) were combined to obtain quasi-spectral complex permittivity measurements in a non-destructive, non-invasive and label-free measurement regime. Such a device can also be used to characterise colloids and suspensions. It is ideal for multi-sensor fusion, such as an 'electronic tongue' device, because of the excellent chemical and biological compatibility and the ease of development of fluidic functionality on the demonstrated platform.

References

[1] Kaatze U, Pottel R and Wallusch A 1995 *Meas. Sci. Tech.* **6** 1201-7
[2] Misra D K 1977 *IEEE Trans. Microw. Theory Tech.* **35** 925-8
[3] Marre S and Jensen K F 2010 *Chem. Soc. Rev.* **39** 1183-202
[4] Kumar V, Paraschivoiu, M and Nigam K D P 2011 *Chem. Eng. Sci.* **66** 1329-73
[5] Santarelli V, MacDonald J and Pine J 1967 *J. Chem. Phys.* **46** 2367-75
[6] Xu M, Eyring E and Petrucci S 1997 *J. Molec. Liq.* **73-4** 41-8
[7] Barthel J, Bachhuber K, Buchner R, Gill J B and Kleebauer M 1990 *Chem. Phys. Lett.* **167** 62-6
[8] Barthel J, Bachhuber K, Buchner R and Hetzenauer H 1990 *Chem. Phys. Lett.* **165** 369-73
[9] Gregory A and Clarke R 2006 *IEEE Trans. Dielectr. Elect. Insul.* **13** 727-43
[10] Pozar D M 2005 *Microwave Engineering 3rd Ed.* (New York: Wiley) chapter 6

Moisture effect on the dielectric response and space charge behaviour of mineral oil impregnated paper insulation

Jian Hao[1, 2], **George Chen**[2], **Ruijin Liao**[1]
[1] The State Key Laboratory of Power Transmission Equipment & System Security and New Technology, Chongqing University, China
[2] School of Electronics and Computer Science, University of Southampton, UK

E-mail: jh210v@ecs.soton.ac.uk

Abstract. Dielectric response and space charge behaviour of oil-paper insulation sample with different moisture contents were investigated using the frequency dielectric spectroscopy (FDS) and the pulsed electroacoustic (PEA) technique, respectively. The influence of moisture on the dielectric response and space charge behaviour of oil impregnated paper insulation was analysed. Results show that the moisture has great effect on the FDS and space charge behaviour of oil impregnated paper insulation. In the frequency range of 10^{-2}~10^{6}Hz, the conductivity and the capacitance of oil impregnated paper increases with its moisture content. The space charge distribution of oil-paper sample with lower and higher moisture contents is very different from each other. The higher the moisture concentration of the oil impregnated paper, the easier the negative charge penetration into the insulation paper. There is a significant amount of positive charge accumulated at the paper-paper interface near to the cathode for oil-paper sample with lower moisture content. However, the positive charge appears in the middle layer paper for oil-paper sample with higher moisture content. Due to the high conductivity, the charge trapped in the oil-paper sample with higher moisture content disappears much faster than that in the oil-paper sample with lower moisture content after removing the voltage.

1. Introduction

Oil/paper insulation system has been widely used in transformers and cables for quite a long time. However, the oil-paper insulation degrades under a combined stress of thermal, electrical, mechanical, and chemical stresses during routine operation [1-2]. Ageing of the oil/paper insulation system produces moisture [1-2]. At the beginning operation of a transformer, the moisture content is usually < 0.5%, but it can reach 4%–8% at the end of its life [3]. The hazard of water in transformers has been given attention in many studies [4-6]. Moisture is recognized to be the “enemy number one” for transformer insulation after the temperature [7]. The higher the water content in the oil-paper system, the more quickly the cellulose degrades. Moreover, moisture is detrimental to the dielectric properties of oil-paper insulation system.

The space charge measurements in oil impregnated cellulose insulation are much more complicated than that in polymers [8]. There are only a few studies that have been performed on oil/paper insulation at the present. Morshuis and Jeroense [9, 10] made some space charge tests on oil impregnated insulation paper, and concluded from the test results that whatever the combination of paper and oil, three general trends were observed. Firstly, homo-charge is always observed, both at the anode and the cathode. Secondly, specific charge growth/decay pattern occurs. Thirdly, the growth and decay of the charge profiles can be described by roughly two time constants. Ciobanu et al [11], analyzed the space charge evolution in oil-paper insulation, and associated the mobility phenomena with the mineral content of cable paper, which is helpful to promote the criteria for the best choice of oil-paper insulation technology for dc cables application. Tang et al [12] investigate the space charge formation in oil/paper insulation system by the PEA method and found that homo-charge injection takes place in multi-layer paper impregnated with oil when subjected to DC electric field. Charge dynamics change with both electric field strength and temperature. The interface between paper layers acts as a barrier that blocks the charge movement. Rongsheng Liu et al [8] studied the distributions of electric field and space charge in oil impregnated pressboard using the pressure wave propagation (PWP) technique at different moisture under DC voltages. It has been found that the moisture plays a very important role in the distribution of space charge in oil-impregnated pressboard. The higher the moisture concentration, the faster is the establishment of space charge. However, the current measured can be affected by polarization and variation of permittivity versus the position of samples.

Taking into account that the moisture has great effect on the insulation life and reliability of the oil-paper insulation system, and in practical application, oil/paper is usually used as multi-layer insulation, in this paper, we have studied the effect of moisture on the FDS and space charge behaviour of three layer oil-paper insulation sample. Firstly, the dielectric property of three layer oil-paper insulation sample with different moisture content was measured. Secondly, space charge dynamics of oil-paper sample with different moisture content were observed using the PEA technique under different DC voltages (4 and 6 kV) at room temperature. The volts-on, volts-off and decay behaviours have been analyzed.

2. Experiments

The insulation oil used in this experiment was Gemini X mineral oil, which was provided by Nynas Oil Company. Gemini X is inhibited insulating oil with extremely good electrical and excellent ageing properties meeting IEC 60296 (03). The new cellulose insulation paper used was provided by ABB Chongqing Transformer Co. Ltd. The single layer thickness of the insulation paper is 75 μm. The parameters of the paper meet the international standard IEC 641-3-1.

Before the PEA tests on oil-paper insulation sample with different moisture contents, several pre-treatment were carried out upon paper samples. Firstly, the new insulation paper was cut into round shape with a diameter of 35mm. A lot of paper samples were kept in the vacuum oven at 130°C for 1 hour for drying. After that, some dried paper samples were impregnated by oil at vacuum condition. And some dried paper samples were put in the air for moisture absorption. After absorbing the moisture for some time, the paper samples were also impregnated by oil under vacuum condition. All the dried paper samples and wet paper samples were impregnated by oil under vacuum condition for one day. Then the moisture content of the dried oil-paper samples and wet oil-paper samples was measurement using Karl Fischer Titration method. The capacitance and conductivity of the oil-paper sample with different moisture contents was measured by dielectric spectroscopy equipment S1260 Impedance/Gain-Phase Analyzer. The dielectric analyzer using a plate-plate geometry electrode with a diameter of 30 mm. Data were obtained over a frequency range of 10^{-2}Hz~10^{6}Hz.

The principle of the PEA can be seen in many literatures [12-14]. The PEA system (PEANUT, made by Five Lab) used in this study has a pulse width of 5ns. The bottom electrode is made of 10mm thick aluminium plate and the top electrode is semiconducting polymer to achieve a better acoustic match. The piezoelectric sensor used was a 9μm thick $LiNbO_3$ material. For each sample, a suitable signal measured at 2 kV was selected as its calibration signal [12, 15].

In practical application, oil-paper is usually used as multi-layer insulation. Therefore, in this experiment, the PEA tests were performed on three layers oil/paper samples (~210μm after oil immersed and being pressed by electrodes, as shown in Figure 1. The samples were stressed at two dc voltage levels (4kVand 6kV) at 15°C. Each time, an electrical stress time of 1 hour was tested. The space charge measurements were taken at various times during the periods of both volts-on and volts-off (short-circuit condition) using the PEA technique. In addition, space charge evolution after the removal of the electric field was also measured.

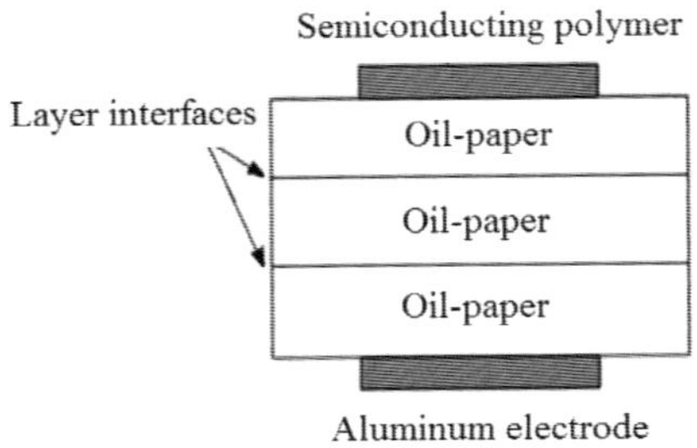

Figure 1. Schematic diagram of sample arrangement

3. Results and Discussions

3.1. FDS of oil-paper sample with different moisture content

As presented in Figure 2, moisture has significant impact on dielectric properties of oil-paper sample. The conductivity of oil impregnated paper increases significantly with increasing moisture content in the frequency range of 10^{-2}~10^{3}Hz. However, there is little difference between the conductivity curves corresponding to different moisture contents at frequencies higher than 10^{3}Hz. The capacitance of oil impregnated paper also increases significantly with increasing moisture content in the frequency range of 10^{-2}~10^{6}Hz, especially in the frequency lower than 10^{2}Hz.

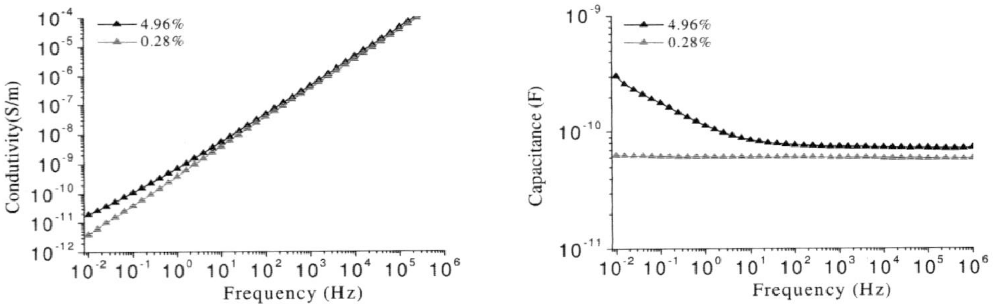

(a) conductivity (b) capacitance

Figure 2. Conductivity and capacitance of the oil impregnated paper

3.2. Space charge behavior of oil-paper sample with different moisture contents

3.2.1 Volts-on

Figure 3 shows the space charge distribution of oil-paper sample with two moisture level under 4kV. Charge injection takes place quickly after voltage is switched on. Homo-charges injection occurs at both electrodes, which leads to the quantities of induced charges on both electrodes decreases as the stressing time increases. Positive charges accumulate in the vicinity of the anode, while the negative charges adjacent to the cathode. The cathode peak is sharp and evident. In contrast, the anode peak is wide and flat. Theoretically, in a PEA figure, the integral area of positive peaks should be the same as that of negative peaks. However, because of the severer acoustic scattering in oil/paper sample, the test results are somehow different from that from that predicted by the theory [12].

For oil-paper sample with 0.28% moisture content, the space charge distribution under 4kV and 6kV is nearly the same. An approximate equilibrium between positive and negative charge injection was observed after 30 min, no significant difference noticed after that time. Along with charge injection, the cathode electrode peaks move towards the inner sample direction. There is a significant amount of positive charge accumulated at the paper-paper interface near to the cathode. Its amount nearly increases with the duration of voltage application. However, the peak position of the positive charge accumulated remains the same in the duration of the voltage application, which further validates the positive charge accumulate at the paper-paper interface. There is also a small amount of negative charge in the middle layer of the oil/paper sample. Its amount also increases with the duration of the voltage application. The presence of both positive and negative charges in the sample suggests that the injection actually occurs at both electrodes. Charge distribution is also complicated between the anode and the paper-paper interface near to the anode due to the severer deterioration of acoustic signal in this region. Compared with 4kV, the relative stable charge density peak of negative charges at the cathode, and the charge density peak value of positive charges accumulated at the paper-paper interface near to the cathode under 6kV is much larger, as shown in Figure 3.

For oil-paper sample with 4.96% moisture content under 4kV and 6kV, the charge injection phenomenon is very different from the oil-paper sample with 0.28% moisture content. Compared with oil-paper with 0.28% moisture content, charge injects more quickly after voltage is switched on under 4kV and 6kV for oil-paper sample with 4.96% moisture content. A lot of negative charge injection occurs at the cathode, which leads to the quantities of negative charges on the cathode electrode decreases fast as the stressing time increases. There is a positive charge peak in the middle layer paper, and the peak value decreases with the voltage application under 6kV obviously. During the voltage application, there is also a negative peak that appears in the paper near to the cathode due to the significant negative charge injection. Compared with 4kV, the charge density peak of negative charges at the cathode and the positive charges accumulated in the middle layer paper under 6kV is larger, as shown in Figure 3(b) and Figure 3(d). Besides, there is mainly negative charge in the oil-paper sample under 6kV at the end of the voltage application.

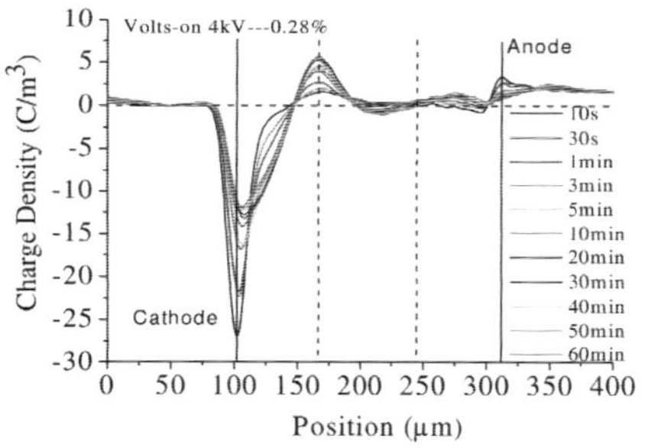

(a) 4kV-0.28%

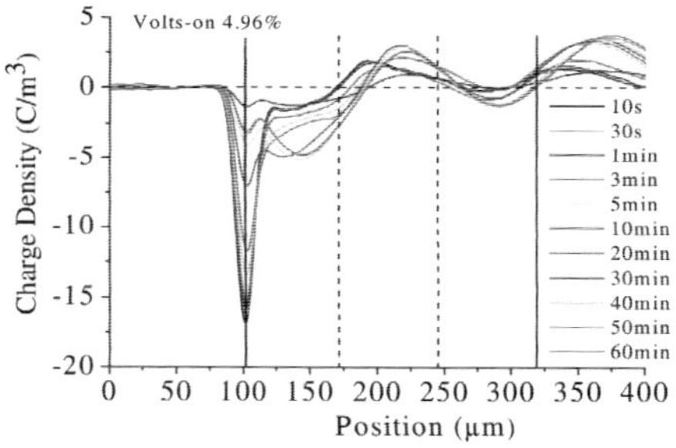

(b) 4kV-4.96%

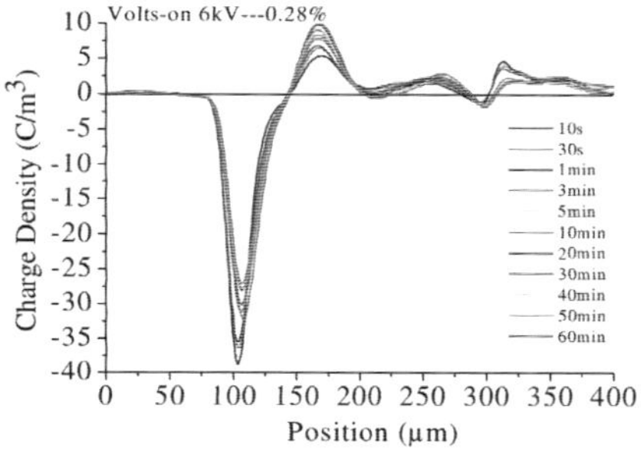

(c) 6kV-0.28%

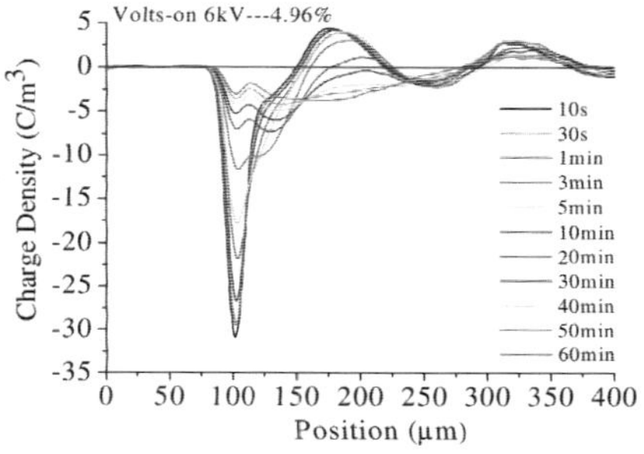

(d) 6kV-4.96%

Figure 3. Volts-on space charge distribution of oil-paper sample with different moisture content

3.2.2 Volts-off

As depicted in publication [12, 15], charges can be roughly classified as fast and slow charges. 'Slow' charges are those which are essentially trapped in deep traps while 'fast' charges are those which escape from the traps very shortly after the DC voltage is removed. The volts-on measurement includes contributions from both fast and slow charges, while the volts-off measurement involves mainly slow charges. Taking into account charges trapped in the insulating material are often stable, i.e. slow charge, in this paper, the volts-off measurement on oil-paper sample with different moisture content had been done, as shown in Figure 4. The distribution of the charge trapped in the oil-paper sample with different moisture contents is different.

For oil-paper sample with 0.28% moisture content under 4kV and 6kV, homo-charges were trapped in the vicinity of both electrodes. Positive charge trapped obviously at the paper-paper interface near to the cathode. In the middle layer paper of the oil/paper sample, apart from the positive charge, there is a small amount of negative charge trapped in the middle position of the single layer paper, and in the single layer paper near to the anode, there is mainly positive charge trapped. Compared to the results under 4kV, the trapped charge distribution under 6kV stress is similar except that the magnitude is higher.

For oil-paper sample with 4.96% moisture content, negative charge trapped in the whole layer paper near to the cathode under 4kV. While there are negative charges trapped in the two layers paper near to the cathode under 6kV, especially at the end of voltage application due to the significant negative charge injection under higher voltage and higher moisture content. Under 4kV, in the middle layer paper and the layer paper near to the anode there is mainly positive charge trapped. While in the layer paper near to the anode there is mainly positive charge trapped under 6kV. The density peak of charges trapped in the oil-paper sample with 4.96% moisture content is lower than that in the oil-paper sample with 0.28% moisture content at the same testing time under 4kV and 6kV. This maybe related to the high conductivity of oil/paper sample with high moisture content.

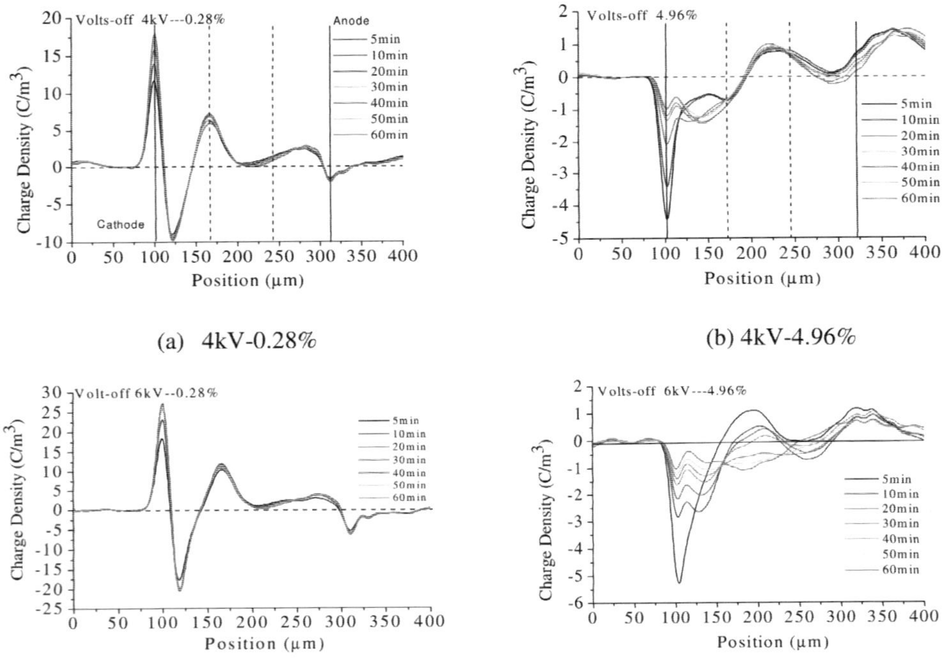

(a) 4kV-0.28% (b) 4kV-4.96%

(c) 6kV-0.28% (d) 6kV-4.96%

Figure 4. Volts-off space charge distribution of oil-paper sample with different moisture content

3.2.3 Decay

After 1 hour of DC voltage applied, space charge distributions of oil-paper sample with different moisture contents after the removal of the applied voltage were measured and are shown in Figure 5. For oil-paper sample with 0.28% moisture content under 4kV and 6kV, the charge decreases with time. After 30 minutes the majority of charge in the bulk including the charge at the interface diminishes through either recombination with the positive charge or conducting away from the sample.

At the initial of decay, as shown in Figure 5(a) and Figure 5(c), it is obvious that the higher of the applied voltage, the larger maximum charge density value of negative charge and positive charge accumulated near to the cathode at the initial stage of decay. While for oil-paper sample with 4.96% moisture content under 4kV and 6kV, the charge trapped in the oil-paper sample disappears within 10s due to the much high conductivity.

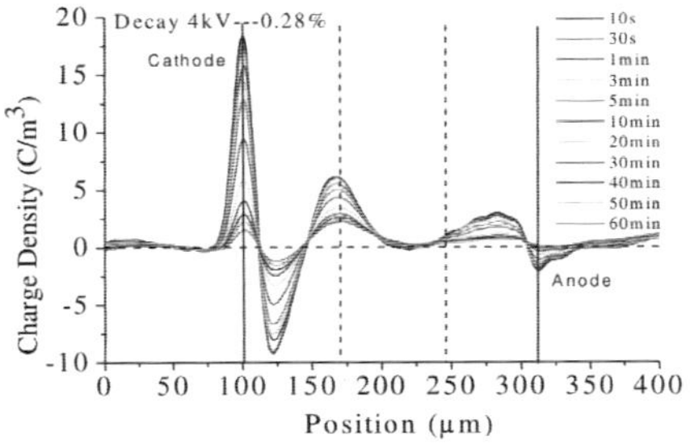

(a) 4kV-0.28%

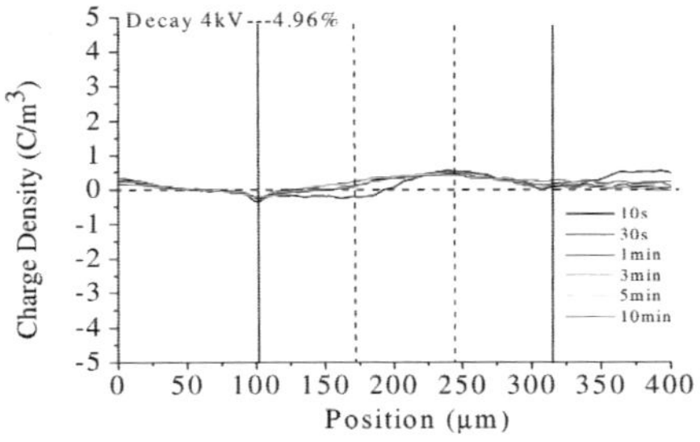

(b) 4kV-4.96%

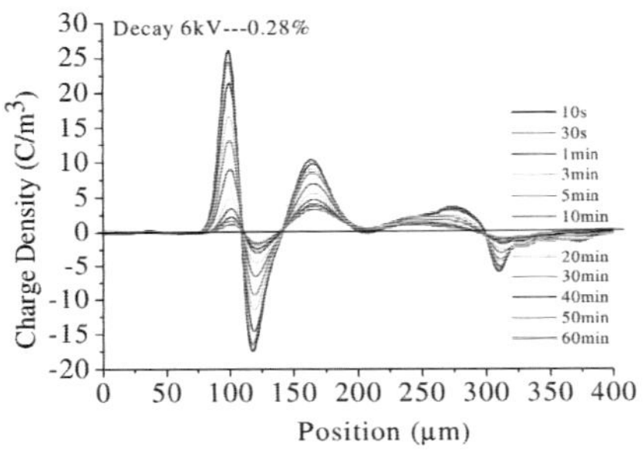

(c) 6kV-0.28%

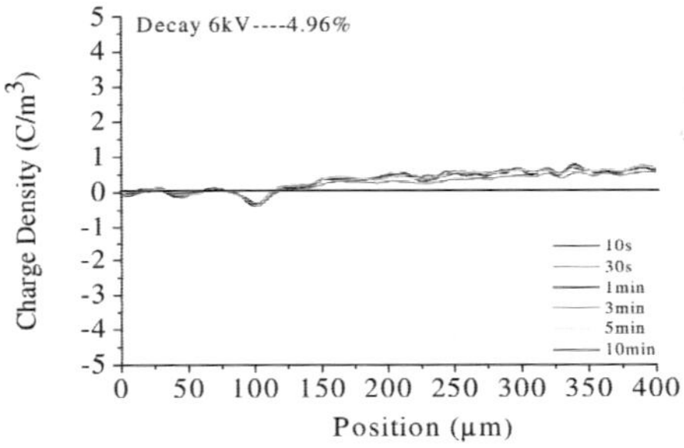

(d) 6kV-4.96%

Figure 5. Decay space charge distribution of oil-paper sample with different moisture content

4. Conclusions

Moisture has significant impact on dielectric properties of oil-paper sample. The conductivity and the capacitance of oil impregnated paper increases significantly with increasing moisture content in the low frequency range.

Homo-charges injection occurs at both electrodes. Charge injected more quickly after voltage was switched on under 4kV and 6kV for oil-paper sample with 4.96% moisture content compared with 0.28% moisture content. There is a significant amount of positive charge accumulated in the paper-paper interface near to the cathode for oil-paper sample with 0.28% moisture content. There is only a positive charge peak in the middle layer paper for oil-paper sample with 4.96% moisture content.

For oil-paper sample with 0.28% moisture content, homo-charges are trapped in the vicinity of both electrodes. Positive charge trapped obviously at the paper-paper interface near to the cathode. While for oil-paper sample with 4.96% moisture content, negative charge trapped in the whole layer paper near to the cathode under 4kV and 6kV. The density peak of charges trapped in the oil-paper sample with 4.96% moisture content is lower than that in the oil-paper sample with 0.28% moisture content at the same testing time under 4kV and 6kV during volts-off measurement.

Due to the much high conductivity, the charge trapped in the oil-paper sample with 4.96% moisture content disappears much faster than that in the oil-paper sample with 0.28% moisture content after removing the voltage.

5. References

[1] Emsley M and Stevens G C 1994 *Cellulose.* **1** 26
[2] Saha T K and Purkait P 2004 *IEEE Trans. Dielectr. Electr. Insul.* **11** 144-154
[3] Emsley A M, Stevens G C 2004 *IEE P-Sci. Meas. Tech.* **141** 324

[4] García B, Burgos J C, ángel A et al. 2005 *IEEE T. Power. Deliver.* **20** 1417
[5] Michael W 1996 The Institution of Electrical Engineers. IEE, Savoy Place, London UK
[6] Du Y, Zahn M, Lesieutm B C et al. 1999 *IEEE Electr. Insul. Mag.* **15** 11
[7] Fofana I, Wasserberg V, Borsi H, Gockenbach E 2002 *IEEE Electrical Insulation Magazine* **18** 18
[8] Rongsheng Liu, Albert Jaksts.1998 *IEEE International Conference on Condudion and Breakdown in Solid Dielectrics,*
[9] Morshuis and Jeroense M 2003 *IEEE Electr. Insul. Mag.* **13** 26
[10] Morshuis P, Jeroense M 1997 *IEEE Conf. Electr. Insul. Dielectr. Phenomena (CEIDP), Minneapolis, USA,* **1** 28
[11] Ciobanu R, Schreiner C, Pfeiffer W and Baraboi B 2002 *IEEE 14th Intern. Conf. Dielectr. Liquids (ICDL), Massachusetts, USA,* 321
[12] Chao Tang, Chen G, Fu M and Liao Ruijin 2010 IEEE *Transactions on Dielectrics and Electrical Insulation* **17** 775
[13] Chen G, Fu M, Liu Z X and Zhong L S 2005 *J. Appl. Phys.* **97** 1-7.
[14] Chen G, Tanaka Y, Takada T and Zhong L 2005 *IEEE Trans. Dielectr. Electr. Insul.* **11** 113
[15] Chen G, Chong Y L and Fu M 2006 *Measurement Science and Technology* **17** 1974

Acknowledgements

All the authors thank the Scientific Research Foundation of SKL of Power Transmission and System Security (2007DA10512708103) for the financial support provided. One of the authors (Jian Hao) wishes to thank the 'Chinese Scholarship Council Funding for Joint Training PhD Student.

Hexagonal boron nitride studied by terahertz time-domain spectroscopy

Mira Naftaly[1], Jon Leist[2], Richard Dudley[1]

[1] National Physical Laboratory, Hampton Road, Teddington, TW11 0LW, UK
[2] Momentive Performance Materials, Inc., 22557 West Lunn Road, Strongsville, Ohio, 44070, USA

mira.naftaly@npl.co.uk

Abstract. Six different grades of hexagonal boron nitride are investigated by terahertz time-domain spectroscopy. The refractive indices and loss coefficients at terahertz frequencies are measured and are related to aspects of material fabrication and properties. Loss is shown to be due primarily to scattering. Porosity is seen to be segregated at the edges of the *c*-planes.

1. Introduction

Boron Nitride is a synthetic ceramic with several possible morphologies, one of which is a hexagonal form with a layered graphite-like structure, as shown in figure 1, commonly termed *h*-BN [1,2]. The *h*-BN structure consists of alternating B and N atoms in the hexagonal planes and also alternating B and N atoms along the crystallographic *c*-axis, forming $(BN)_3$ rings with very strong intra-layer bonding and weak van der Waals bonding between layers. There is a large difference in the separation of atoms in the layers (1.45 Å) and between the planes (3.33 Å) [3].

Hot-pressing methods have been developed to achieve nearly full-density shapes of *h*-BN, as well as BN composites with other ceramic materials. A comprehensive list of properties of *h*-BN is available at [2] http://www.ioffe.rssi.ru/SVA/NSM/Semicond/BN/basic.html. Although *h*-BN has good transparency in the far-infrared, there is little published data [1,2,4-6], and no attempt has been made to relate far-infrared transmission to aspects of material properties.

In this paper we investigate four grades of *h*-BN using THz time-domain spectroscopy. The variations in the refractive index and loss are discussed in relation to material fabrication and structure. The utility of h-BN as an optical material at THz frequencies is also explored.

2. Material fabrication and properties

All boron nitride specimens were manufactured by Momentive Performance Materials and are representative of commercially available forms. Table 1 lists the types of BN and their properties.

Due to the low surface energy of the *c*-plane, BN does not self-sinter [7]. High-density freestanding shapes are achievable by hot pressing submicron, turbostratic BN in the presence of a binder phase under temperatures approaching 2000 °C and pressures up to 2000 psi. BN/AlN composites are manufactured using similar techniques. The selection of binder phase is dictated by the final application requirements, and may be boric oxide (B_2O_3) or calcium borate ($Ca_3(BO_3)_2$). The binder phase may be leached from the shape after it is generated. Pyrolytic BN is manufactured by vacuum deposition from a gas phase at temperatures of 1800-1900 °C and pressures below 1 Torr.

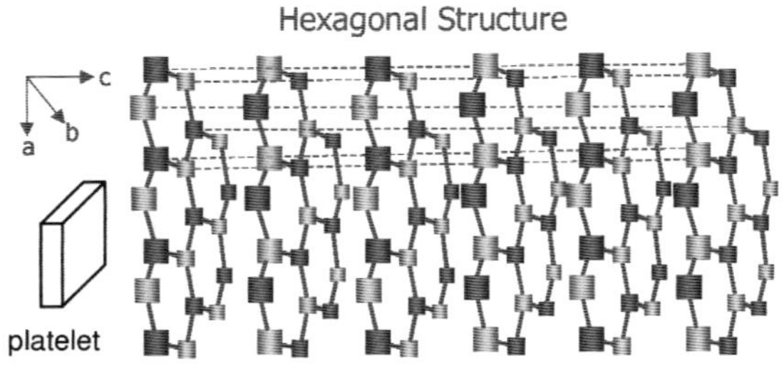

Figure 1. BN structure showing crystallographic planes in relation to crystal growth.

During fabrication, the pressed plug is formed as an array of standing platelets, which assume a preferential orientation with respect to the pressing direction such that their *c*-plane is approximately parallel to the pressing axis. Each platelet consists of hexagonal crystals aligned with their *ab*-plane lying in the plane of the platelet. Materials having hexagonal crystal structure are expected to be optically anisotropic, the resulting birefringence being such that the extraordinary ray lies in the *ab*-plane and the ordinary ray is aligned with the *c*-axis (figure 1).

Grades PBN, HBC and HBT consist of pure BN. Grades HBN and HBR contain binders, while BIN77 is a BN/AlN composite. PBN is fully dense; other grades contain varying fractions of porosity.

3. Data acquisition and analysis

The THz time-domain spectrometer (TDS) used a standard configuration incorporating a femtosecond laser, four off-axis parabolic mirrors, a biased GaAs emitter, and electro-optic detection with a ZnTe crystal and balanced photodiodes. The maximum dynamic range of the system was 5000 in amplitude, and the frequency resolution in the experiments was 7.5 GHz. The samples were placed in the collimated part of the THz beam, which had a diameter of 25 mm and was vertically polarised. Measurements were carried out in dry air in order to eliminate water absorption lines from the recorded spectra. The amplitude and phase of the THz signal as a function of frequency are obtained from the measured time-domain data of THz electric field by applying the Fourier Transform using a standard FFT application (OriginPro 8). The experimental setup and procedures are described in more detail elsewhere [8].

Loss coefficients (α) and refractive indices (n) of the samples were measured by comparing THz transmission through two different thicknesses of the sample material, and were calculated from the transmitted spectra by using the equations [8]:

$$\alpha(\nu) = -\frac{2}{d_1 - d_2} \ln\left[\frac{E_2(\nu)}{E_1(\nu)}\right] \quad (1)$$

$$n(\nu) = 1 + \frac{c[\phi_2(\nu) - \phi_1(\nu)]}{2\pi\nu(d_1 - d_2)} \quad (2)$$

where $E_{1,2}(\nu)$ and $\phi_{1,2}(\nu)$ are the amplitude and phase of the THz field at the frequency ν, and $d_{1,2}$ are the sample thicknesses. Loss coefficients $\alpha(\nu)$ combine contributions from absorption and scattering.

4. Results

The refractive indices and loss coefficients of the *h*-BN grades examined are summarised in table 1. Figure 2 shows an example of data for the grades HBC. The results for HBN and HBR are similar to those in [4] (where the corresponding grades are A and HP). The THz optical properties of different BN grades may be analysed in relation to their microscopic structures.

All grades with the exception of BIN77 are birefringent with negative birefringence. The weak positive dispersion in all samples is attributed to the tail of the lowest phonon resonance at 23 THz [9].

Grade BIN77 is non-birefrigent because it is isotropic on the macroscopic scale, being formed as a mixture of randomly oriented crystals of BN and AlN. The refractive index of AlN is 2.92 [10]; while the mean index of fully dense BN is 2.11 (see PBN). The refractive index of BIN77 is 2.53, and is therefore consistent with its composition of 50BN:50AlN (mol%) and 10% porosity.

Grade PBN is a fully dense, high-purity, highly oriented material, so its refractive index values lie closest to those of a single-crystal: its n_e is lower than that of sintered grades, and its n_o is higher, due to the more uniform and precise alignment of its crystal planes, resulting in higher birefringence.

It may be expected that the refractive indices of different BN grades will decrease linearly with increasing porosity, since less material lies in the optical path. Figure 3 depicts the relationship, where PBN is seen to lie outside the trend of other grades. The HBR grade has refractive index values below the trend, attributed to the presence of ~6% of $Ca_3(BO_3)_2$ as binder. The other binder-containing grade, HBN, conforms to the trend, because B_2O_3 forms a glassy matrix with a similar refractive index to BN. The most notable feature of figure 3 is that the n_e decreases with porosity, as expected, but the n_o does not. This supports the view that porosity in these materials is located predominantly at the edges of the *c*-planes, and not at the low-surface-energy faces of *c*-planes. The *e*-ray is polarised in the plane of the platelets, therefore experiencing an effective refractive index which is modified (reduced) by the interstices between the platelet edges. The *o*-ray, which is polarised transversely to the platelets, should be similarly affected by interstices between platelet layers. Consequently, the fact that the n_o remains constant indicates the absence of porosity between platelet faces.

Loss in ceramic materials is caused by combined contributions of absorption and scattering. Single-crystal *h*-BN has high transparency at THz frequencies [5,6], therefore losses in different BN grades, listed in table 2, are due primarily to scattering, with some contribution from binder absorption. Scattering loss is expected to rise with porosity due to the greater presence of scattering centres. As expected, loss in fully-dense PBN is much lower than in other grades. Indeed, PBN is sufficiently transparent to be a good candidate material for THz optical components. In contrast, BIN77 is much more lossy than other grades, although like BN, AlN is highly transparent [10]. The loss is attributed to increased scattering arising from the mismatch of the refractive indices of the two constituents (2.1 for BN and 2.9 for AlN). Loss in HBC, HBN and HBT indeed increases with porosity; whilst HBR has higher loss due to absorption by $Ca_3(BO_3)_2$. HBT is of interest because of its pronounced dichroism. This is attributed to porosity being segregated at the edges of the platelets, thus causing greater scattering of the beam polarised in the *ab*-plane.

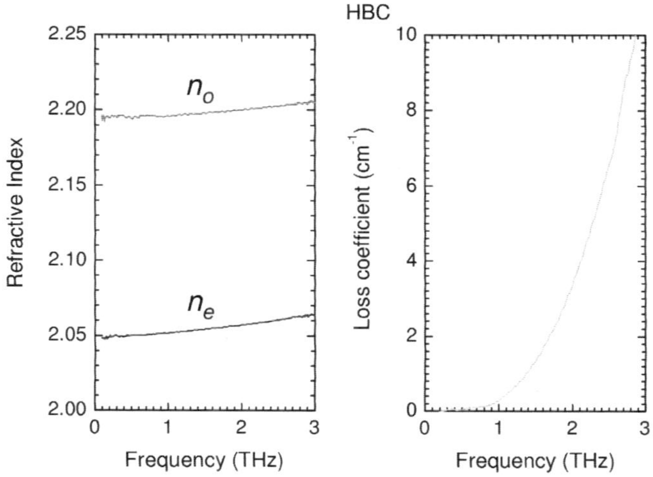

Figure 2. THz refractive indices and loss coefficient of the HBC grade of boron nitride.

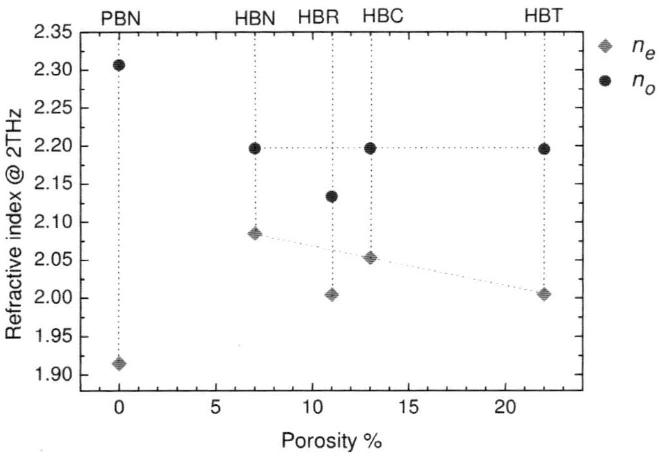

Figure 3. THz refractive indices of BN grades at 2 THz.

Table 1. Properties of different grades of Boron Nitride.

	PBN	HBN	HBR	HBC	HBT	BIN77
BN mol%	100	>95	>94	>99	>99	47
AlN mol%	-	-	-	-	-	47
Binder	None	B_2O_3	$Ca_3(BO_3)_2$	None	None	$Ca_3(BO_3)_2$
Density (g/cm^3)	2.26	2.10	2.00	1.95	1.75	2.43
Porosity %	0	7	11	13	22	10
$n_e \pm 0.002$ @ 2THz	1.915	2.087	1.995	2.058	2.015	2.530
$n_o \pm 0.002$ @ 2THz	2.307	2.197	2.128	2.200	2.202	2.530
n_e- n_o @ 2THz	-0.392	-0.110	-0.133	-0.142	-0.187	0
Loss ± 5% (cm^{-1}) @ 2THz	0.85	2.2	10.6	3.0	13.1 (*e*) 5.2 (*o*)	70

5. Conclusions

Six grades of ceramic BN were examined and THz optical properties were shown to be related to the material structure. Porosity was seen to be segregated at the edges of the *c*-planes, and loss was seen to be due to scattering. The effects of different binders and BN/AlN composite were analyzed.

Acknowledgements
Work at NPL was supported by the National Measurement Office.

References

[1] Geick R, Perry C H and Rupprecht G 1996 *Phys. Rev.* **146** 543
[2] http://www.ioffe.rssi.ru/SVA/NSM/Semicond/BN/basic.html
[3] Moore A W 1990 *J. Cryst. Growth* **106** 6
[4] Gatesman A J, Giles R H and Waldman J 1993 *J. Appl. Phys.***73** 3962
[5] Ishii T and Sato T 1983 *J. Cryst. Growth* **61** 689
[6] Gielisse P J, Mitra S S, Plendl J N, Griffis R D, Mansur L C, Marshall R and Pascoe E A 1967 *Phys. Rev.* **155** 1039
[7] Rusanova L N and Gorchakova L I 1989 *Powder Metal. Metal Ceram.* **28** 108
[8] Naftaly M, Leist J, Dudley R A 2010 *J. Eur. Ceram. Soc.* **30** 2691
[9] Hoffman D M, Doll GL, Eklund P C 1984 *Phys. Rev. B* **30** 6051
[10] Naftaly M, Greenslade PJ, Miles R E, Evans D 2009 *Opt. Mater.* **31** 1575

Influence of copper on the by-products of different oil-paper insulations

Jian Hao[1, 2], Ruijin Liao[1], George Chen[2] and Chao Ma[3]
[1] The State Key Laboratory of Power Transmission Equipment & System Security and New Technology, Chongqing University, China
[2] School of Electronics and Computer Science, University of Southampton, UK
[3] Gansu Electric Power Research Institute, China

E-mail: cquhaojian@126.com

Abstract. Transformer failure caused by the corrosion of copper material in transformer attracts great attention of researchers and engineers. In this paper, Karamay No. 25 naphthenic mineral oil, Karamay No. 25 paraffinic mineral oil, Kraft paper and copper were used to compose four combinations of oil-paper insulation samples. The ageing by-products and dielectric properties of the four combinations of oil-paper insulation samples were compared after they were thermally aged at 130°C. The influence of copper on the by-products and dielectric properties of different oil-paper insulations was obtained. The results show that copper can accelerate the ageing rate of insulation oils and reduce their AC breakdown voltage. The content of copper substance dissolved in insulating oil increases with ageing time at first and then decreases. The paper aged in the oil-paper insulation sample with copper has higher moisture content than the one without copper. Results of energy dispersive spectroscopy (EDS) in the scanning electron microscope (SEM) show that there is copper product deposited on the surface of insulation paper. The insulation oil and paper aged in the oil-paper insulation sample with copper have higher dielectric loss and conductivity than that without copper.

1. Introduction

Transformers play an important role in providing a reliable and efficient electricity supply and are one of the most critical equipments in electric power transmission and distribution systems. The insulation system in most power transformers consists of oil/paper on the copper windings, and there are also several oil-impregnated pressboard barriers between the high and low voltage windings, and sometimes also between windings and the core [1]. Copper has stable physico-chemical properties, low electrical resistivity and is widely used in power transformers [2]. At present, there are many investigations about the corrosive sulphur compounds present/formed in the insulating oil reacted with the copper windings and other copper parts [3-11], research on the influence of copper corrosion on the ageing characteristics of oil-paper insulation during its ageing process is limited.

In this paper, accelerated thermal ageing experiment of four different combination oil-paper insulation samples at 130°C, where observations on naphthenic oil-paper insulation and paraffinic oil-paper insulation samples both with and without copper were compared, respectively. Oil parameter includes the oil acidity, AC breakdown voltage and the copper substance content in the oils. Paper parameter has been carried out including paper colour, SEM and the moisture content of paper. At last, the influence of copper on the dielectric properties of oil-paper insulation sample was also provided.

2. Experiments

The test materials are common Kraft paper, Karamay No. 25 naphthenic mineral oil, Karamay No. 25 paraffinic mineral oil and rectangular copper. Insulating oil used in the experiment was provided by China Chongqing Chuan Run Co. Ltd. The parameters of naphthenic oil and paraffinic oil meeting the specifications of IEC 60296 are presented in Table 1. The abbreviations of four combination oil-paper insulation samples are presented in Table 2. Rectangular copper strips, whose dimensions were 5 mm in width, 7.5 mm in length and 0.2 mm in thickness, were used. They were polished with sand paper firstly to obtain active surfaces. After polishing, the strips were immersed in naphthenic oil and paraffinic oil respectively to avoid the oxidation of surface. Kraft paper, whose dimensions were 5mm in width and 0.07 mm in thickness, were wrapped five turns around the strip and then the bundle was bound with cord.

Table 1. Typical parameters of naphthenic and paraffinic oil

Property	Naphthenic oil	Paraffinic oil
Density at 20°C (kg/m^3)	882.9	872.8
Kinematic viscosity 40°C (mm^2/s)	9.655	10.06
Pour point (°C)	-45	-24
Flash point (°C)	140	159
Acidity (mg KOH/g)	0.01	0.022
Breakdown voltage (2.5mm gap electrodes) (kV)	55	42
Corrosive sulphur	non-corrosive	non-corrosive

Table 2. Samples used for thermal ageing test

Sample name	Sample composition
NK	naphthenic oil + Kraft paper
NKC	naphthenic oil + Kraft paper + copper
PK	paraffinic oil + Kraft paper
PKC	paraffinic oil + Kraft paper + copper

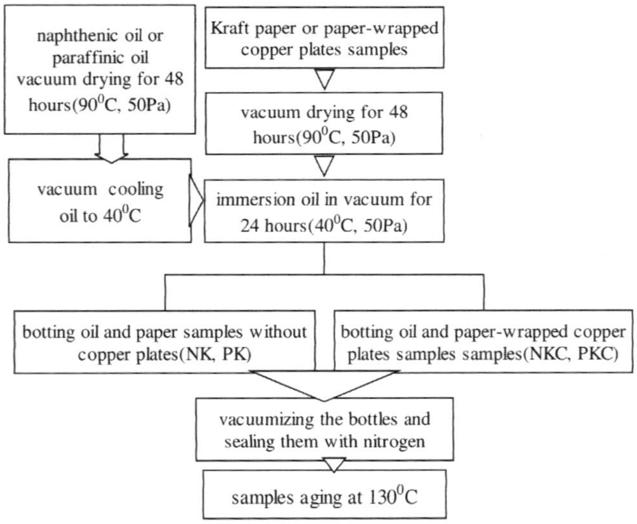

Figure 1. Pre-treatment of samples and the flow chart of ageing test

The pre-treatment of samples and the flow chart of ageing were shown in Figure 1. After being pre-treated, the moisture content of paper is 0.70%. The oil samples have a moisture content of 9 mg/kg. 400 ml oil and the paper (7.5 g) or paper-wrapped copper strip (paper: 7.5 g) were placed in the same glass vial (500 ml), respectively, as presented in Figure 1. For each sample of NK, NKC, PK and PKC, there are 8 vials. The space above the oil was filled with nitrogen and the vial was sealed tightly. After that, all vials were placed in a temperature controlled bath at 130°C with a temperature deviation of 0.1°C for accelerated thermal ageing. Figure 2 shows the photographs of the sample prepared well. The parameters of oil and paper were analyzed after ageing different times.

Figure 2. Photographs of initial ageing samples

3. Results and Discussions

3.1. Oil acidity

As a direct parameter to reflect the ageing degree of insulation oil, the oil acidity of NK, NKC, PK and PKC samples are shown in Figure 3. The oil acidity was measured according to ASTM D974-04. The acidity values presented in this paper is the average of three testing values for each sample at room temperature 27±0.2 °C. The accuracy of the measurement is about ±0.005 mg KOH/g. Oxidation of oils can be accelerated by acids and catalysed by copper substance, such as copper ion and heavy copper-organic compounds [12, 13]. It can be seen from Figure 3 that NKC samples with copper have higher oil acidity than NK samples without copper all the time. However, PKC samples with copper have lower oil acidity than PK samples without copper before ageing 29 days, after which the oil acidity of PKC samples become higher than PK samples. Besides, the oil acidity of NK and NKC is lower than oil acidity of PK and PKC after ageing the same time at 130 °C, respectively. It can be seen from Figure 3 that Copper play a catalyse role on the whole ageing of naphthenic oil, while the catalyse role becomes obvious in paraffinic oil after ageing 29 days.

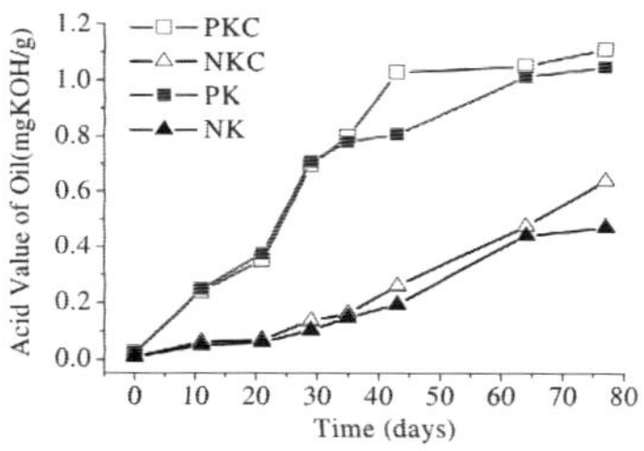

Figure 3. Changes of oil acidity during oil-paper insulation samples ageing process

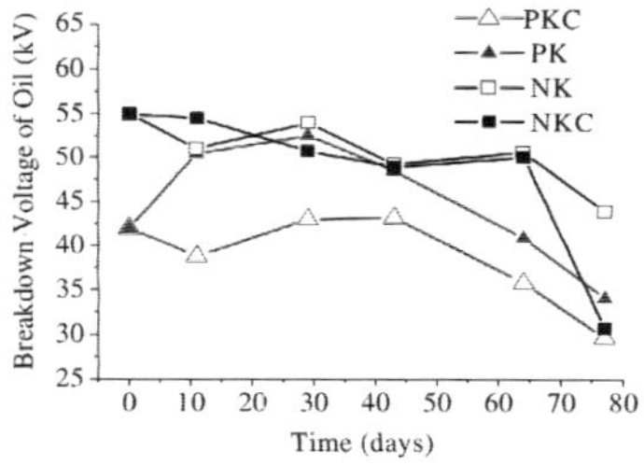

Figure 4. Breakdown voltage of insulation oils versus ageing time

3.2. AC breakdown voltage of oil

The most important and therefore most often checked parameter of a liquid insulation is the breakdown voltage. The AC breakdown voltage of oils in NK, NKC, PK and PKC samples as function of the ageing time is shown in Figure 4. The AC breakdown voltage of oil was measured according to IEC 60156. The AC breakdown voltage value presented in this paper is the average of six testing values for each sample at room temperature 27±0.2 °C. The accuracy of the measurement is 1.5%. The AC breakdown voltage of oil in NK sample is higher than oil in NKC sample during whole ageing process, except for the sample aged 11 days. The AC breakdown voltage of oil in PK sample is also higher than oil in PKC sample during all the ageing process. It indicates that copper substance existence in oil-paper insulation during the ageing process weaken the oil dielectric property. In addition, regardless of oil-paper insulation sample with copper or without copper, Figure 4 shows that the AC breakdown voltage of naphthenic oil is nearly all higher than that of paraffinic oil during the thermal ageing process. This presents that the naphthenic oil performs a better dielectric characteristic than paraffinic oil at the same ageing condition.

3.3. The content of copper substance dissolved in oil

Copper substance such as copper ion and heavy copper-organic compounds play an important role in the ageing process of oil-paper insulation, which is shown above. The oil acidity of paraffinic oil is much higher than naphthenic oil during the whole ageing process (Figure 3). The increased oil acidity will increase the corrosion rate of copper. Therefore, the copper corrosion in PKC sample is more serious than that in NKC sample, as presented in Figure 5. Figure 5 shows that the content of copper substance dissolved in paraffinic oil is larger than that in naphthenic oil before ageing 35 days. However, when sampled at 35 days, the sludge was observed in all aged samples, which is more obvious in PK and PKC sample. The copper-organic compounds in the oil will sediment together with the sludge and hence decrease the copper content dissolved in oil [12, 14, 15]. The more formation of insoluble sludge will cause more heavy copper-organic compounds sediment in the sludge. Consequently, a decreased of the content of copper substance dissolved in oil is observed after ageing 35 days, and the content of copper substance dissolved in paraffinic oil is less than that in naphthenic oil.

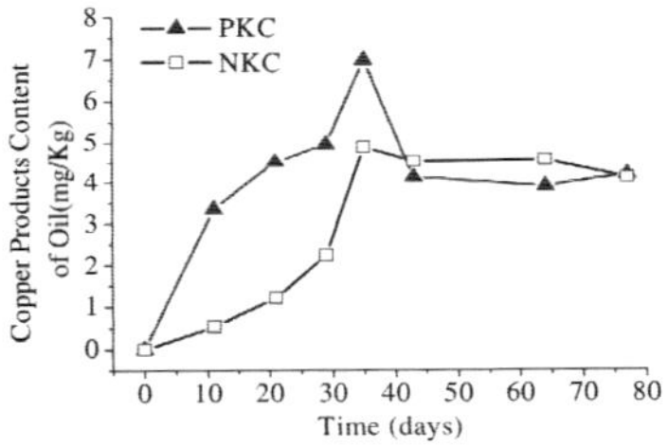

Figure 5. Content of copper substance dissolved in oil

3.4. Moisture content of paper

The absolute moisture content of oil impregnated paper of NK, NKC, PK and PKC sample is shown in Figure 6. Moisture moves between the paper and insulation oil to reach equilibrium in terms of relative saturation. Therefore, the moisture content of oil impregnated paper of NK, NKC, PK and PKC sample is undulate. It can be seen from Figure 6 that the copper has great effect on the moisture content of oil impregnated paper in the ageing process. The PK and PKC sample nearly has higher moisture content than the NK and NKC sample in the whole ageing process. It's noteworthy that the moisture content of paper aged in PKC is higher than that aged in PK, and the moisture content of paper aged in NKC is higher than that aged in NK.

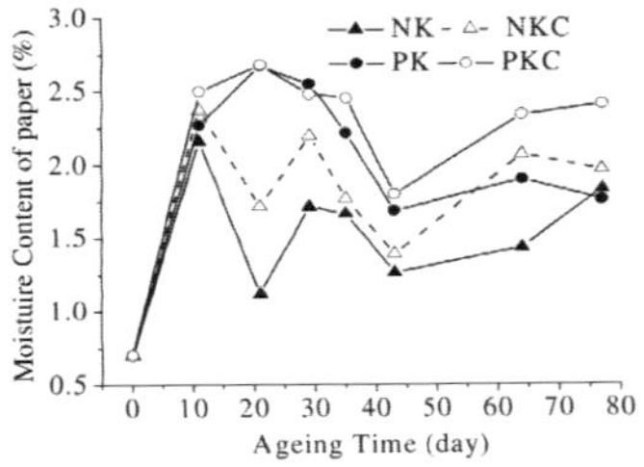

Figure 6. Moisture content in paper varying with ageing time

3.5. Paper property

The difference in appearance between the innermost papers near to the copper plate aged in NK, NKC, PK and PKC sample are shown in Figure 7. The aged paper becomes more brittle with ageing time increased, especially for the paper aged in PK and PKC sample. Visually, the longer time of the paper aged, the more discoloration of the paper became. After ageing 77 days, the paper in paraffinic oil was almost black, while that in naphthenic oil was yellow. These observations appear to predict a faster degradation rate of paper in paraffinic oil than in naphthenic oil.

Compared with the paper aged in NK and PK, it should be emphasized that that there is obvious product deposition phenomenon in NKC and PKC sample when sampling at 77 days. Copper deposition in the insulating paper has been widely reported in recent years [3-11] and it is copper dissolution by corrosion that subsequently leads to copper deposition on insulating paper. From a chemical perspective, copper can, in principle, be deposited in the form of oxides, sulphides, or other compounds and complexes, or even as metallic copper after redox reactions [12]. The innermost paper near to the copper plate of NKC samples aged for 77 days were measured by SEM. Before the SEM measurement performed on the paper sample, the paper was gold coated. Figure 8 shows that there is copper element deposited on the surface of paper.

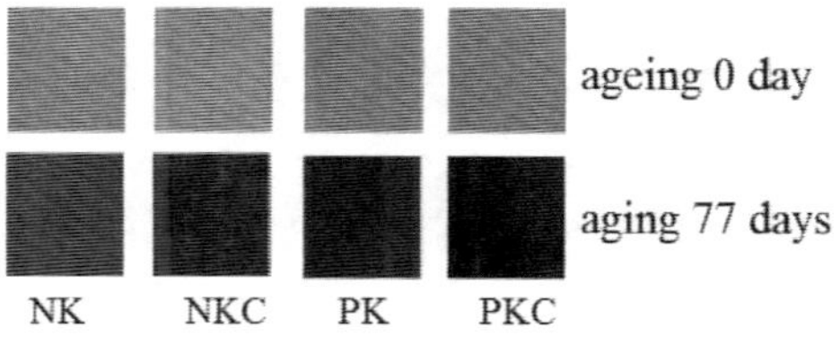

Figure 7. Colour of the innermost paper near to the copper plate in NK, NKC, PK and PKC sample

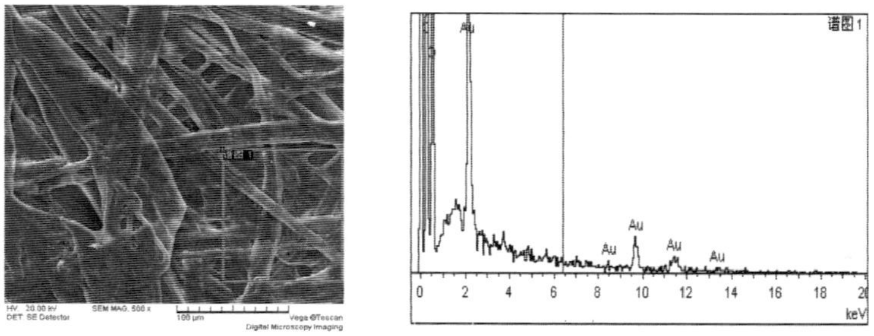

(a) The innermost paper near to the copper plate in NK aged for 77 days

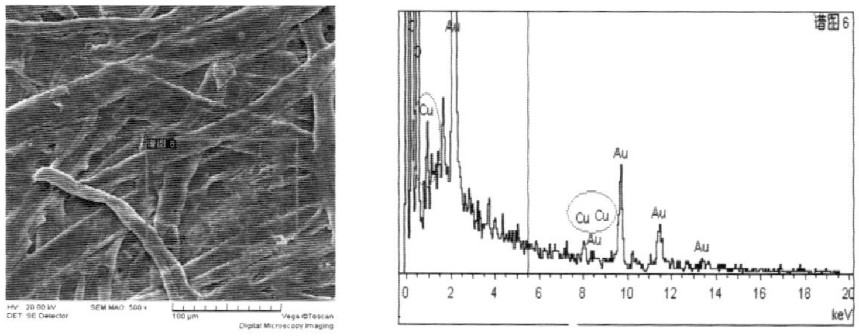

(b) The innermost paper near to the copper plate in NKC aged for 77 days

Figure 8. SEM/EDS result of innermost paper near to the copper plate in NK and NKC aged for 77 days

4. Conclusions

In this work, thermal ageing test of four combinations of oil-paper insulation samples including naphthenic oil + Kraft paper (NK), naphthenic oil +Kraft paper + copper (NKC), paraffinic oil + Kraft paper (PK) and paraffinic oil + Kraft paper + copper (PKC) at 130 °C has been carried out, where observations on naphthenic oil-paper insulation and paraffinic oil-paper insulation samples both with and without copper were compared, respectively. The following conclusions are able to be drawn:

NKC samples with copper have higher oil acidity than NK samples without copper all the time. However, PKC samples with copper have lower oil acidity than PK samples without copper before ageing 29 days, after then the oil acidity of PKC samples become higher than PK samples. Copper play a catalyse role on the whole ageing of naphthenic oil, while the catalyse role becomes obvious in paraffinic oil after ageing 29 days.

Copper substance existence in oil-paper insulation during the ageing process weakens the oil AC breakdown. Naphthenic oil performs a better AC breakdown characteristic than paraffinic oil at the same ageing condition.

The content of copper substance dissolved in paraffinic oil is larger than that in naphthenic oil before ageing 35 days, while after then, a decreased of the content of copper substance dissolved in oil is observed, and the content of copper substance dissolved in paraffinic oil is less than that in naphthenic oil.

Copper has great effect on the moisture content of oil impregnated paper in the ageing process. SEM results show that there is copper product deposited on the insulating paper surface.

5. References

1. Linhjell D, Lundgaard L E, Gafvert U 2007 *IEEE Transactions on Dielectrics and Electrical Insulation* **14** 156
2. Wood-Mallock J C, Wood L G 1958 *J. Inst. Pet.* **44** 320
3. CIGRE WG A2-32. 2007 Copper sulphide in transformer insulation. *ELECTRA*, No. 230,: 12
4. CIGRE TF A2.31. 2006 Copper sulphide in transformer insulation. *ELECTRA*, No. 224, 20
5. CIGRE WG A2-32. Copper sulphide in transformer insulation. Interim Report No.1, 2006.
6. Bengtsson C, Dahlund M, Hajek J. et al. *CIGRE*, Paris, 2006.
7. ABINEE, Workshop on Corrosive Sulphur in Oil, June 2-3 2005, Sao Paulo, Brazil. Available: http://www.cigre-a2.org/Site/Publications.
8. Kobayashi T, Tanimura J, Murakami H, et al. 2007 *CIGRE, SC A2& D1*, Colloquium, Belgium.
9. Eechhoudt S, Van J. Peteghem, Kinable L 2007 *CIGRE, SC A2& D1*, Colloquium, Belgium, PS2-05,
10. Bastos G M. 2007 *CIGRE, SC A2& D1*, Colloquium, Belgium, PS0-01,
11. Scatiggio F, Tumiatti V, Maina R, et al. 2008 *IEEE Trans. Power Delivery* **23** 508
12. Wiklund P, Levin M, Pahlavanpour B 2007 *IEEE Electrical Insulation Magazine* **23** 6
13. Liao R J, Tang C, Yang L J, et al. *IET Electr. Power Appl.* 2009 **3** 407
14. Thomson C N1958 J. Inst. Petrol. **44** 295
15. Kalantar A, Levin M 2008 *Lubrication Science* **20** 223

Acknowledgements
All the authors thank the Scientific Research Foundation of SKL of Power Transmission and System Security (2007DA10512708103) for the financial support provided.

On the effect of nanosilica on a polyethylene system

K Y Lau[1, 2, 3], A S Vaughan[1], G Chen[1] and I L Hosier[1]

[1] School of Electronics and Computer Science, University of Southampton, Southampton SO17 1BJ, United Kingdom
[2] Faculty of Electrical Engineering, Universiti Teknologi Malaysia, 81310 Johor Bahru, Malaysia

E-mail: kyl1g10@ecs.soton.ac.uk

Abstract. The topic of polymer nanocomposites remains an active area of research in terms of its potential applications in dielectric and electrical insulation applications. Although more than a decade has passed since Lewis first considered these systems as dielectric materials, the precise effects of incorporating nanofillers into different polymers are yet to be confirmed. While this strategy is generally regarded as a promising means for tailoring insulation performance, the consequences of adding a nanofiller to an insulating matrix are not always positive; for example, where dispersion is poor the breakdown strength can be adversely affected. This paper reports on an investigation into a nanofilled polyethylene system prepared via a solution blending route. Specifically, in this study, the effect of processing and nanofiller dispersion on the structure and physical properties of the nanocomposites that forms was considered. The crystallization behaviour and morphology of the differently processed materials have been evaluated by differential scanning calorimetry (DSC) and polarised optical microscopy (POM). The influence of nanofiller dispersion on breakdown behaviour is also described.

1. Introduction

Polymer nanocomposites, defined as polymers in which nanometre-sized fillers are homogeneously dispersed at just a few weight percentage (wt%), remains as an interesting topic of discussion in the context of dielectrics and electrical insulation. Since Lewis [1] first considered these systems as dielectric materials, numerous publications have emerged, exploring the dielectric properties of various polymers combined with different inorganic nanofillers [2-7]. While there were numerous results supporting the dielectric enhancement brought about by nanostructuration in electrical insulation systems, such results were not always positive. Furthermore, lots of uncertainties concerning nanocomposite applications in dielectric systems remain unanswered and are yet to be confirmed.

This paper reports on an investigation into a nanofilled polyethylene system prepared via a solution blending route. This involves the initial dissolution of the polyethylene in nonpolar xylene and the dispersion of the nanosilica in relatively polar methanol – a non-solvent for polyethylene. Mixing together the two components results in the rapid precipitation of the polymer, including the nanoparticles. The use of solution blending method is not new in nanodielectrics research and has been

[3] To whom any correspondence should be addressed.

explored by some researchers [8-11]. This preparation method enables many different formulations to be produced with ease and, as such, is ideal for a scoping survey, providing a different perspective on the understanding of nanodielectrics.

2. Experimental

2.1. Materials and Preparation

The polymers used in this research were low density polyethylene (LDPE) grade LD100BW obtained from ExxonMobil Chemicals and high density polyethylene (HDPE) grade Rigidex HD5813EA obtained from BP Chemicals. The nanofiller used was silicon dioxide (SiO_2) nanopowder (i.e. nanosilica) obtained from Sigma-Aldrich, with a quoted particle size range from 10 nm to 20 nm.

The desired amount of nanosilica was added into polar methanol – a non-solvent for polyethylene and dispersed through sonication. Concurrently, the proper amount of HDPE and LDPE were dissolved in xylene under heating and stirring. The hot xylene/polyethylene mixture was then poured into the methanol/nanosilica mixture quickly with vigorous stirring, resulting in the rapid precipitation of the nanocomposite as a white mass. Upon filtering and drying, the resulting nanocomposite was melt pressed at 150 °C and vacuum dried at 100 °C. Samples for different tests were subsequently prepared by melt pressing at a temperature of 150 °C, followed by direct quenching into water or isothermal crystallisation. For comparison purpose, the unfilled polyethylene system was prepared in the same way.

For convenience, all materials were denoted using the following general notation: P/F/T. In this, P signifies the polymer used, F represents the filler used and T defines the thermal treatment used. Examples of designations are summarised as in Table 1.

Table 1. Materials investigated.

Designation	Composition (wt%)			Description
	Polymer		Filler	
	LDPE	HDPE	Nanosilica	
PEA/0/Q	80	20	0	Polyethylene system type A (i.e. the one with 20 wt% of HDPE and 80 wt% of LDPE), not added with nanofiller, and being quenched directly into water.
PEA/5S1M/Q	80	20	5	Polyethylene system type A (i.e. the one with 20 wt% of HDPE and 80 wt% of LDPE), added with 5 wt% of nanosilica being sonicated for 1 hour in methanol, and being quenched directly into water.
PEA/0/115	80	20	0	Polyethylene system type A (i.e. the one with 20 wt% of HDPE and 80 wt% of LDPE), not added with nanofiller, and being isothermally crystallised at 115 °C.
PEA/5S1M/115	80	20	5	Polyethylene system type A (i.e. the one with 20 wt% of HDPE and 80 wt% of LDPE), added with 5 wt% of nanosilica being sonicated for 1 hour in methanol, and being isothermally crystallised at 115 °C.

2.2. Characterisation

A Perkin Elmer DSC7 with Perkin Elmer Pyris software was used to characterise the thermal behaviour of the materials being investigated. For each differential scanning calorimeter (DSC) measurement, a sample of ~5 mg in weight was used, placed in a sealed aluminium pan. The experiment was performed in a nitrogen atmosphere at a scan rate of 10 °C/min. Polarised optical microscopy (POM) was used to evaluate the morphology of the materials.

Dielectric breakdown tests were conducted based upon the general consideration laid down in the ASTM Standard D149-87. The test sample, with a thickness of about 85 μm, was placed between two 6.3 mm diameter steel ball bearings immersed in Dow Corning 200/20cs silicone fluid, with an applied AC voltage of 50 Hz and a ramp rate of 50 V/s until failure. 20 breakdown tests were performed on each type of sample. The breakdown data were statistically analysed using the two-parameter Weibull distribution analysis.

3. Results and Discussion

3.1. Thermal Analysis

The DSC melting behaviour for the materials investigated is shown in Figure 1. In the case of unfilled polyethylene, at least two melting peaks can be observed in both the quenched and isothermally crystallised materials. The lower melting peaks are associated with the LDPE-rich phase, while the upper melting peak is associated with the HDPE component of the blend. Such observation is not unusual and has been reported in other studies [12, 13].

Comparing both the quenched, nanofilled and quenched, unfilled polyethylene systems, the melting behaviour of PEA/5S1M/Q is similar to PEA/0/Q, with a lower peak at ~114 °C and an upper peak at ~125 °C. For the case of the isothermally crystallised systems, the lower and upper melting peaks of the nanofilled material, PEA/5S1M/115 were ~106 °C and ~125 °C, respectively, with no significant difference from the unfilled material, PEA/0/115. Consequently, the DSC thermal traces indicated that there were no thermal changes caused by nanosilica inclusion.

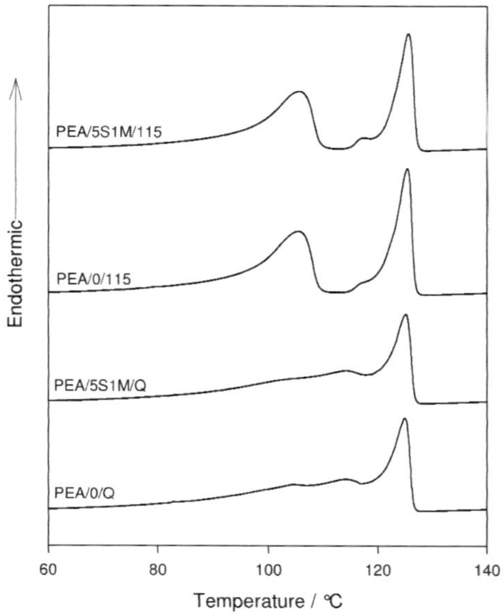

Figure 1. DSC melting traces.

3.2. Morphological Analysis

Figure 2 illustrates the dependence of morphology on thermal treatment and material composition. In the unfilled polyethylene system that had been isothermally crystallised at 115 °C (Figure 2(a)), spherulites can be clearly observed by POM. Isothermal crystallisation at 113 °C (Figure 2(b)) and 117 °C (Figure 2(c)) also show clear evidence of spherulites in this material. However, as shown in Figure 2(d), the nano-inclusion appears dramatically to suppress spherulitic development. In the case of both the unfilled and nanofilled polyethylene systems that had been quenched, no spherulites can be observed by POM, as shown in Figure 2(e) and Figure 2(f), respectively.

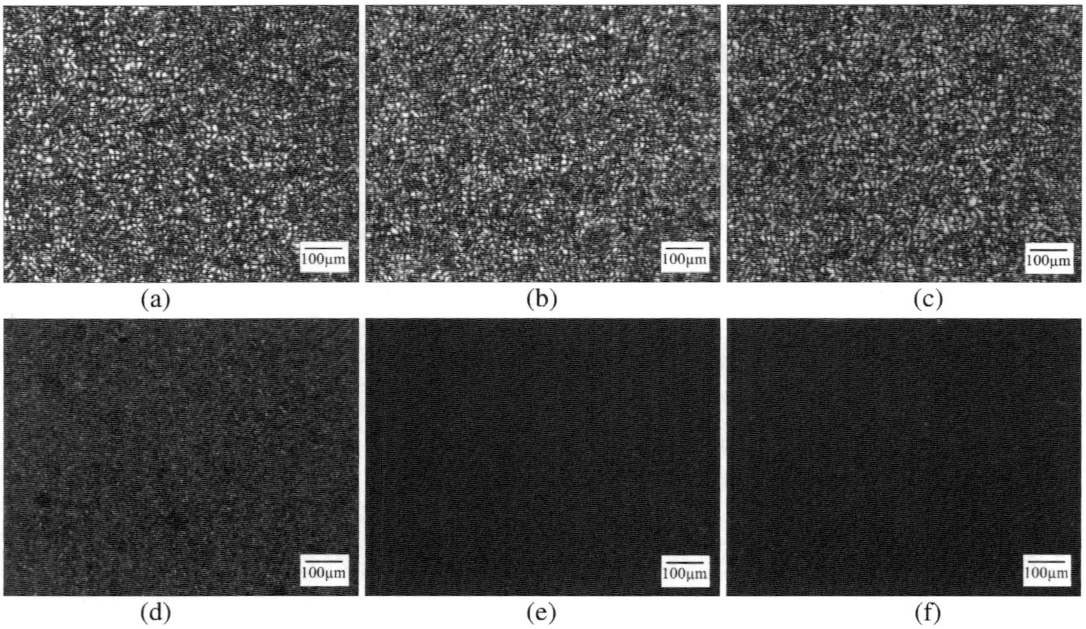

Figure 2. Optical micrographs taken under crossed polarisers for (a) PEA/0/115, (b) PEA/0/113, (c) PEA/0/117, (d) PEA/5S1M/115, (e) PEA/0/Q, (f) PEA/5S1M/Q.

3.3. Breakdown Strength Analysis

Figure 3(a) and Figure 3(b) show Weibull plots that compare the breakdown strength of the quenched and isothermally crystallised materials, respectively. It can be noticed that the breakdown strength of PEA/5S1M/Q and PEA/5S1M/115 were significantly lower than that of PEA/0/Q and PEA/0/115, respectively. In the case of unfilled polyethylene systems, the breakdown fields of PEA/0/Q and PEA/0/115 were about 148 kV/mm and 152 kV/mm, respectively. The addition of nanosilica caused the breakdown field to reduce to about 132 kV/mm and 138 kV/mm for PEA/5S1M/Q and PEA/5S1M/115, respectively. Therefore, in this preparation route, the addition of nanosilica caused a reduced breakdown strength in both quenched and isothermally crystallised polyethylene systems.

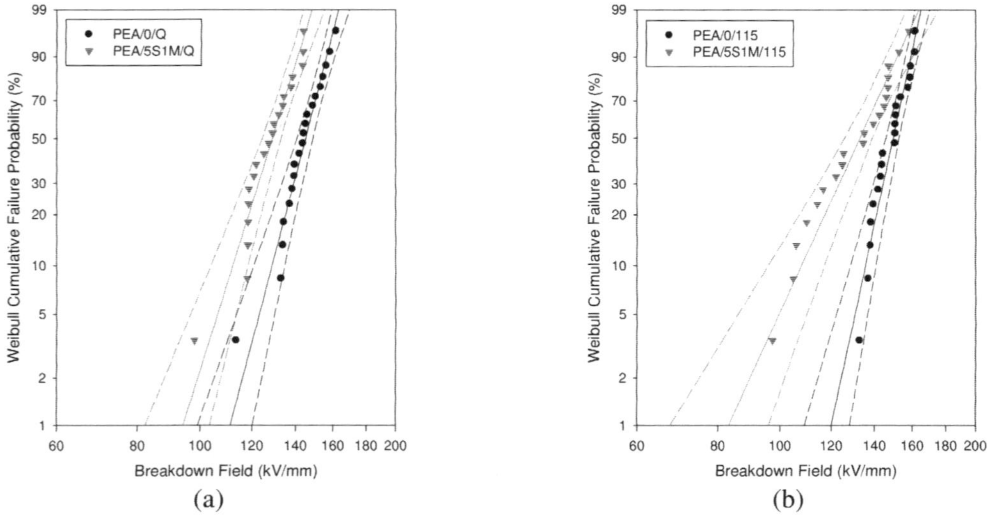

Figure 3. Weibull plots for (a) PEA/5S1M/Q vs. PEA/0/Q, (b) PEA/5S1M/115 vs. PEA/0/115.

The effect of nanoparticle sonication on the breakdown strength was also studied, as shown in Figure 4. For PEA/5S1M/Q and PEA/5S1M/115, the nanosilica was sonicated in methanol for 1 hour prior to precipitation with the polyethylene/xylene mixture. Increasing the sonication duration to 2 hours, with materials denoted as PEA/5S2M/Q and PEA/5S2M/115 for the quenched and crystallised nanofilled polyethylene blends, respectively, did not help in improving the breakdown strength. Table 2 summarises the Weibull parameters obtained through a two-parameter Weibull distribution analysis.

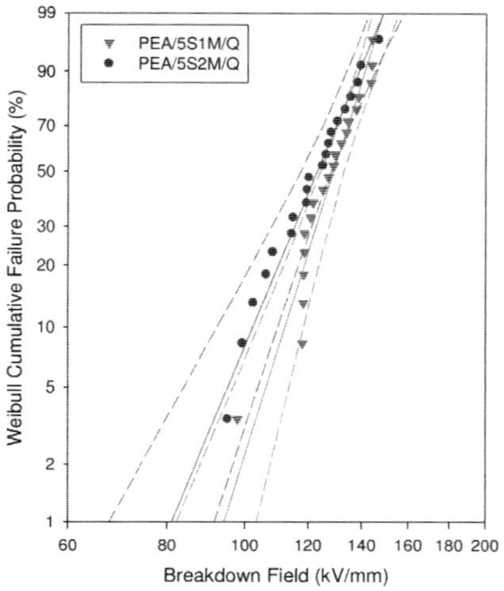

(a)

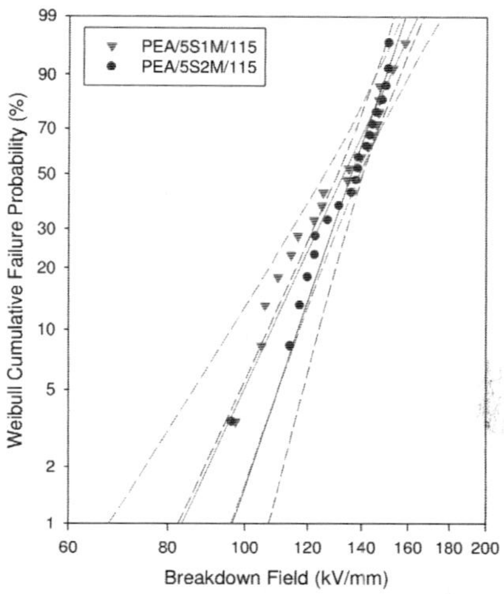

(b)

Figure 4. Weibull plots for (a) PEA/5S1M/Q vs. PEA/5S2M/Q, (b) PEA/5S1M/115 vs. PEA/5S2M/115.

Table 2. Weibull parameters.

Material	Scale parameter (kV/mm)	Shape parameter
PEA/0/Q	148 ± 4	16 ± 5
PEA/5S1M/Q	132 ± 4	13 ± 4
PEA/5S2M/Q	128 ± 5	10 ± 3
PEA/0/115	152 ± 3	19 ± 6
PEA/5S1M/115	138 ± 6	9 ± 3
PEA/5S2M/115	140 ± 5	12 ± 4

4. Conclusions

Nanofilled and unfilled polyethylene systems were prepared by a solution blending route. Considering both the quenched and isothermally crystallised polyethylene systems, the DSC melting behaviour was not altered by nano-inclusion. In the isothermally crystallised polyethylene systems, the dispersion of nanosilica caused less pronounced spherulitic development, as determined by POM. The breakdown strength was significantly reduced due to nano-inclusion, regardless of the nanoparticle sonication duration in methanol. The lowered breakdown strength could be related to the preparation and/or morphology of the materials, but more detailed analysis, such as the use of scanning electron microscopy, is required for such understanding.

5. References

[1] Lewis T J 1994 *IEEE Trans. Dielectr. Electr. Insul.* **1** 812–25
[2] Kozako M, Fuse N, Ohki Y, Okamoto T and Tanaka T 2004 *IEEE Trans. Dielectr. Electr. Insul.* **11** 833–9
[3] Nelson J K, Fothergill J C, Dissado L A and Peasgood W 2002 *Annu. Rep. Conf. Electrical Insulation and Dielectric Phenomena (Cancun)* 295–98.
[4] Roy M, Reed C W, MacCrone R K, Schadler L S, Nelson J K, Keefe R and Zenger W 2005 *Proc. Int. Sym. on Electrical Insulating Materials (Kitakyushu)* 223–6
[5] Tanaka T, Montanari G C and Mülhaupt R 2004 *IEEE Trans. Dielectr. Electr. Insul.* **11** 763–84
[6] Vaughan A S, Swingler S G and Zhang Y 2006 *IEEJ Trans. Fund. Mater.* **126** 1057–63
[7] Zilg C, Kaempfer D, Thomann R, Mülhaupt R and Montanari G C 2003 *Annu. Rep. Conf. Electrical Insulation and Dielectric Phenomena (Albuquerque)* 546–50.
[8] Yin Y, Chen J, Yang J, Xiao D, Tu D, Yin R and Qian H 2003 *Proc. Int. Conf. Properties and Applications of Dielectric Materials (Nagoya)* 913–16.
[9] Huang X Y, Jiang P K and Kim C U 2007 *J. Appl. Phys.* **102** 124103
[10] Huang X, Jiang P, Kim C, Ke Q and Wang G 2008 *Compos. Sci. Technol.* **68** 2134–40
[11] Huang X, Ma Z, Wang Y, Jiang P, Yin Y and Li Z 2009 *J. Appl. Polym. Sci.* **113** 3577–84
[12] Hosier I L, Vaughan A S and Swingler S G 1997 *J. Mater. Sci.* **32** 4523–31
[13] Hosier I L, Vaughan A S and Swingler S G 2010 *J. Mater. Sci.* **45** 2747–59

Acknowledgements

One of the authors (K Y Lau) would like to acknowledge Ministry of Higher Education, Malaysia and Universiti Teknologi Malaysia for the financial sponsorship.

On sample preparation and dielectric breakdown in nanostructured epoxy resins

M Reading, Z Xu, A S Vaughan[1] and P L Lewin
The Tony Davies High Voltage Laboratory, University of Southampton, UK

E-mail: asv@ecs.soton.ac.uk

Abstract. There are many methods available to achieve a good dispersion of fillers within a polymeric matrix. This investigation considered several methods of dispersing three chosen fillers within an epoxy resin; the same processes were also performed on unfilled materials to investigate any effects they may have on the host material. For this investigation, the epoxy system (EP) was combined with sodium montmorillonite (MMT), micrometric silicon dioxide (SD) or nanometric silicon dioxide (NSD) as fillers. The effect of the different sample preparation routes on breakdown behaviour was then evaluated. While more thorough mixing protocols were found to lead to improved breakdown behaviour in the case of the various filled systems, surprisingly, an entirely equivalent form of behaviour was also seen in the unfilled epoxy. The influence of changes in sample geometry on the breakdown strength was established.

1. Introduction

There has been much interest in creating composite dielectrics over recent years, many involving a polymer matrix with a filler or additive present to increase one or more specific properties of the overall material. Fillers were originally, and can still be, used as cheap diluents to reduce costs and improve material production and processing, this being the origin of the term filler [1]. However in recent years, more focus has been given to fillers that actively improve the final-material properties. By dispersing such fillers into a host material, such as a polymer, it has been seen that properties such as electrical breakdown strength, crystallisation kinetics, fire retardancy and more, can be altered to create superior materials for chosen purposes [2]. When creating a polymer composite using a particulate filler, many issues must be addressed, including the compatibility of the filler with the polymer, the size of the filler particulates, the agglomeration of the filler, the filler loading and the dispersion of the filler within the polymer [3]. It is commonly known that a poor dispersion of filler will result in a material with non-uniform properties, often a particularly undesirable attribute [4, 5]. Therefore, the aim for most of these composite systems is for the filler to be as well dispersed as possible, giving uniform behaviour across the whole material. Much work has been performed on fillers for polymers and Rothon [1] provides a good review on the use of such particulate fillers, including those with nanoscopic dimensions.

This study was undertaken with three objectives. The first of these was to investigate methods of dispersing fillers with varying geometries and particle sizes within an epoxy matrix; the second, which

[1] To whom any correspondence should be addressed.

was directly related to the first, involved determining the influence of dispersion on electrical breakdown behaviour. The subsequent data and its analysis will reveal the effectiveness of the dispersion methods for the varying systems and whether the dispersion achieved varies due to the particle geometry and size. The third objective relates to the effect of sample thickness on the measured breakdown strength. It is well known that the apparent breakdown strength is influenced by volume effects (sample thickness and area tested) and, therefore, fluctuations in sample thickness will influence the scatter seen in breakdown data [6]. Also, since our interest in the epoxy-based systems studied here concerns their potential use in full scale plant, it is important to evaluate the extent to which conclusions drawn from thin films can reliably be extrapolated to thicker, technologically relevant geometries.

2. Sample Details and Production

DER 332 epoxy resin was chosen for this investigation, due to its simple chemistry and appropriate viscosity, allowing easy mixing. Jeffamine D-230 was chosen as the hardener based upon previous experience in this laboratory [7]. The standard stoichiometric ratio calculation leads to a resin:hardener ratio of 1000:344. The fillers used for this investigation were a sodium montmorillonite (MMT) formed of platelets nanometric in one dimension and two sizes of nominally spherical particulate silicon dioxide, a micro filler 1-5 μm in size (indicated SD) and a nano filler containing particles 5-15 nm in diameter (indicated NSD). All fillers were obtained from Sigma Aldrich and were used as received.

For electrical breakdown measurements, film specimens of the required thickness (50 – 1000 μm) were produced using a gravity-fed moulding technique. Figures 1 and 2 show one of the moulds used to produce these films. A QZ13 release agent was applied to the moulds prior to assembly since this has previously been effective at aiding release of DER 332 epoxy systems [8]. All epoxy samples, including the filled systems, were given the same curing treatment, with the fillers being added and dispersion methods performed at the de-gassed resin stage.

Figure 1. One of the pre-made sample moulds

Figure 2. The spacer and sample inside the pre-made mould from figure 1

Figure 3 shows a flowchart of the procedure followed in producing the various samples and Table 1 lists the sample types generated during this investigation and the mixing regime performed on each sample. The five sample preparation regimes were employed to produce samples exhibiting dispersion states ranging from poorly dispersed through to a highly dispersed state. Regime NM leads to a poor dispersion state, as the sample is completely unmixed. Regime HM gives a low state of dispersion with only basic hand mixing being used. Regime MS provides an intermediate level of dispersion, with magnetic stirring being performed using a standard heated stirrer plate held at ~50 °C. Regimes US and SON represent attempts at achieving a higher state of dispersion using an ultrasound bath and a sonicator respectively. In regime US an UltraWave DP201 Precision Ultrasonic bath was used at 50 °C while, in regime SON a Hielscher UP200S Ultrasonic processor was used at 50% amplitude.

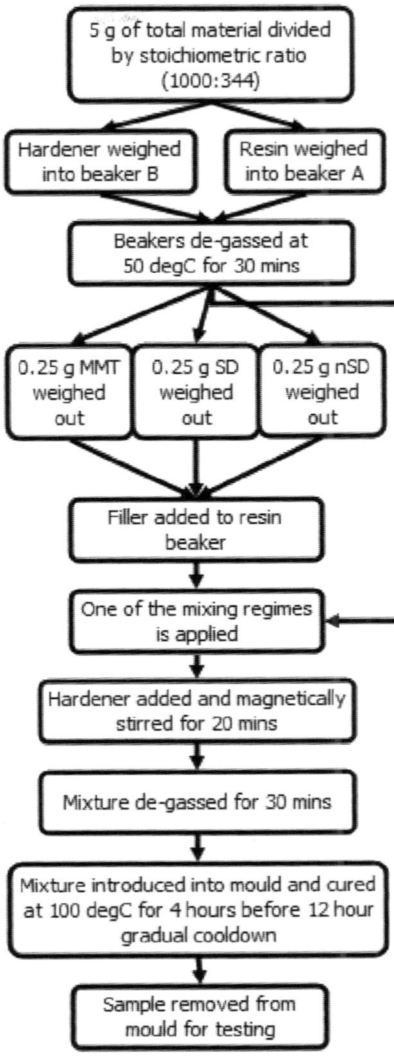

Figure 3. A flowchart showing sample production

Table 1. Samples generated and their corresponding mixing regime

Sample	Filler %wt	Mixing Regime
EP NM	-	NM
EP HM	-	HM
EP MS	-	MS
EP US	-	US
EP SON	-	SON
MMT NM	5% MMT	NM
MMT HM	5% MMT	HM
MMT MS	5% MMT	MS
MMT US	5% MMT	US
MMT SON	5% MMT	SON
SD NM	5% SD	NM
SD HM	5% SD	HM
SD MS	5% SD	MS
SD US	5% SD	US
SD SON	5% SD	SON
NSD NM	5% NSD	NM
NSD HM	5% NSD	HM
NSD MS	5% NSD	MS
NSD US	5% NSD	US
NSD SON	5% NSD	SON

These two techniques rely on the same fundamental technique to break-up aggregations of fillers and improve the dispersion with the polymer, although the delivery method is slightly different since the ultrasound bath exposes the outside of the flask whereas the sonicator is a point source located in the middle of the specimen. Table 2 contains further details of the procedures used.

To examine the effect of specimen thickness, the unfilled epoxy was used in conjunction with mixing regime MS and different spacers were used during sample preparation. For samples between 50 μm and 500 μm, Melinex films (DuPont) were used; 1 mm thick specimens were produced using an aluminium spacer.

3. Experimental

For the study of mixing effects, electrical breakdown data were obtained using an AC ramp testing rig developed in-house [9]. In line with the ASTM D149 standard (AC ramp, 50 Hz, 50 V/s), tests were performed on typically 15 sites per sample. The test cell contained silicone oil, which covered the

Table 2. Details of the five mixing regimes

Regime	Details
NM	Resin, hardener and filler combined with no mixing
HM	Hand mixing for 20 minutes
MS	Hand mixing for 20 minutes followed by magnetic stirring for 20 minutes
US	Hand mixing for 20 minutes, magnetically stirred for 20 minutes and ultrasound for 20 minutes
SON	Hand mixing for 20 minutes, magnetically stirred for 20 minutes and sonicated for 20 minutes

sample to prevent flashover, and 6.3 mm steel ball bearing electrodes were used. The electrodes were changed after every other sample to avoid pitting and the linearity of the AC ramp rate was also checked. The resulting data were analysed using a two parameter Weibull expression with 90% maximum likelihood confidence bounds. Since the maximum reliable operating voltage of the above breakdown rig is limited to about 15 kV, it was incapable of breaking down 1 mm thick specimens. Consequently, the study of thickness effects used, in its place, a Phenix Type 600C AC dielectric test set and an equivalent test protocol to that described above.

4. Results

4.1. Mixing Regime

Figures 4 to 7 present electrical breakdown data obtained from the unfilled and filled composite specimens 100 μm in thickness. Figure 4 shows that even in the absence of any nanofiller, the extent of mixing has a dramatic effect on the data; the scale parameter, α, increases progressively from

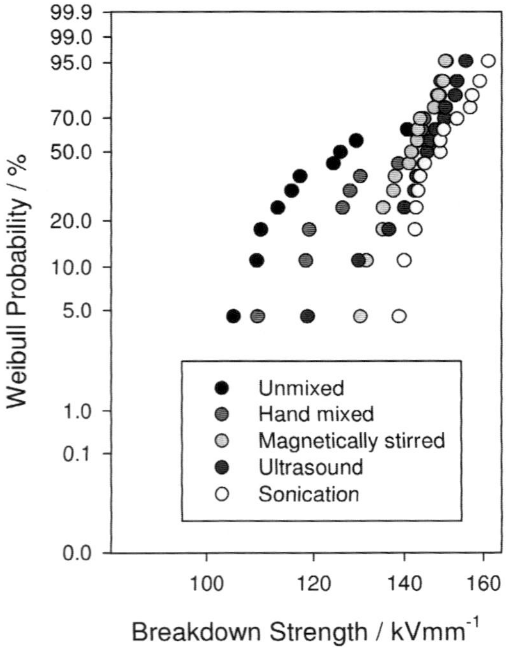

Figure 4. Weibull plots of breakdown data for differently processed unfilled samples

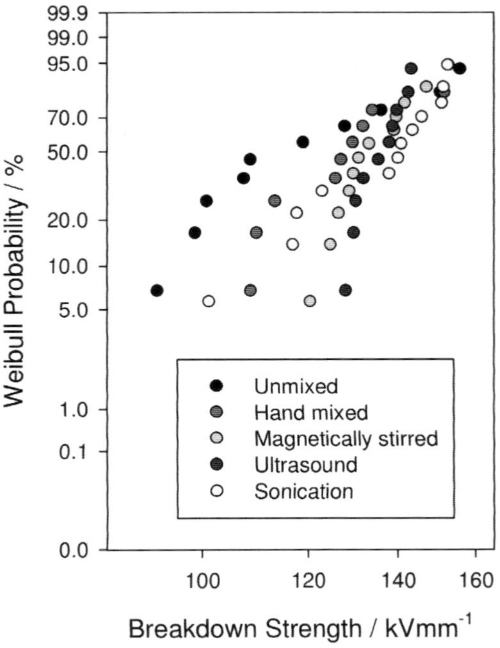

Figure 5. Weibull plots of breakdown data for differently processed MMT-filled samples

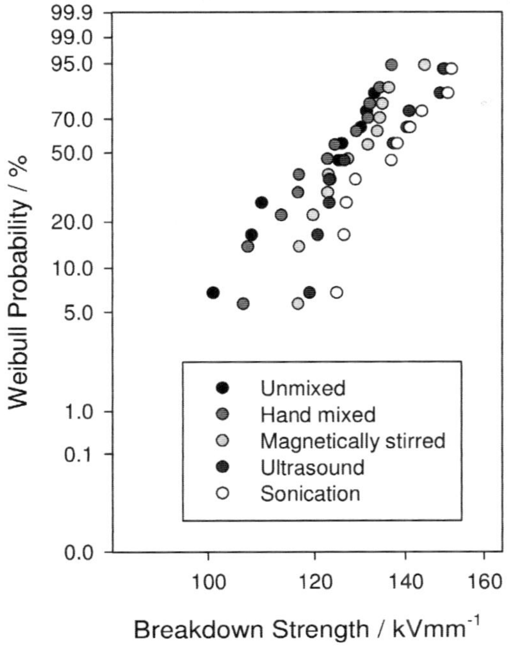

Figure 6. Weibull plots of breakdown data for differently processed SD-filled samples

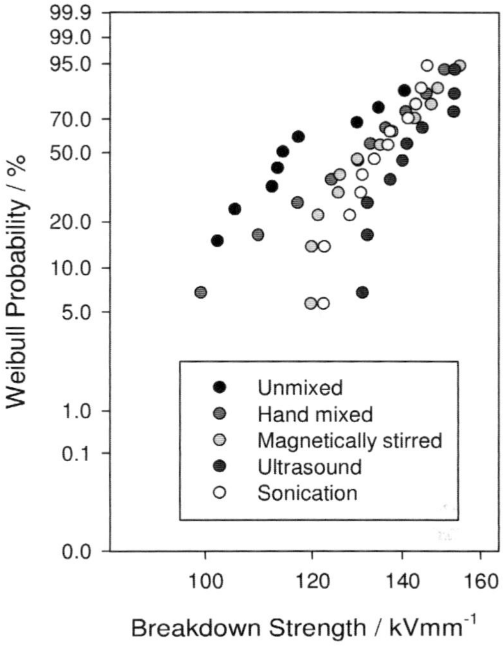

Figure 7. Weibull plots of breakdown data for differently processed NSD-filled samples

~135 kV mm^{-1} to a value in excess of 150 kV mm^{-1} and the shape parameter, β, increases from ~9 to more than 20; the best breakdown performance is obtained from the sonicated system. Although such procedures have been advocated in connection with nanofilled epoxy, we are not aware of any previous accounts showing such a marked effect in the unfilled resin.

Qualitatively equivalent behaviour is also seen in the various filled systems; as more thorough mixing regimes are used, the spread in the data drops and the breakdown strength progressively increases. However, quantitatively, these results are not what we would expect. In filled systems, the

Table 3. The Weibull parameters obtained from the plots shown in figures 4-7

Sample	Weibull Parameter		Sample	Weibull Parameter	
	α	β		α	β
EP NM	135.8	9.1	SD NM	129.8	9.88
EP HM	141.3	14.3	SD HM	127	14.62
EP MS	143.8	26.5	SD MS	132.5	16.2
EP US	147.5	21.9	SD US	138.3	13.1
EP SON	152	22.17	SD SON	141.5	15.7
MMT NM	128.9	6.1	NSD NM	129.2	5.6
MMT HM	137.8	7.52	NSD HM	135.3	10.18
MMT MS	139.2	14.9	NSD MS	139.5	12.3
MMT US	138.5	29.9	NSD US	145.9	18.4
MMT SON	141.8	11.52	NSD SON	138.5	19.9

reduced filler agglomeration that should result from more intensive mixing should augment any benefits seen in the unfilled analogue. This is not seen here. In all three systems, the NM materials exhibit breakdown strengths values close to 129 kV mm^{-1} (c.f. ~136 kV mm^{-1} for the unfilled resin), which subsequently rises for the sonicated samples to ~142 ± 4 kV mm^{-1} (c.f. ~150 ± 2 kV mm^{-1} for the unfilled resin). While more intensive mixing is beneficial in enhancing the breakdown behaviour of nanocomposites, the fact that equivalent behaviour is also seen in the same, unfilled resin, makes it very difficult to argue that this effect is related to improved dispersion of the filler, as is widely claimed in the literature [10, 11].

4.2. Specimen Thickness

For studying new or filled/altered materials, the use of thin films has several benefits; in that the volume of material required is low, as is the voltage range required to cause breakdown. However, even small variations in sample thickness can then have a disproportionate effect on the apparent breakdown strength, leading to increased scatter in the resulting Weibull plots. Figure 8 shows the thickness dependence of breakdown data obtained from unfilled epoxy specimens prepared using mixing regime MS. At a thickness of 50 µm, all samples breakdown at applied fields close to 180 kV mm^{-1} (applied voltage 9 kV) whereas, when the thickness is increased to 1 mm, this falls to ~40 kV mm^{-1} (applied voltage 40 kV). Plotting the derived location parameter as a function of sample thickness yields the graph shown in figure 9. For this system, the thickness dependence of the breakdown strength, E_b, conforms closely to an equation of the form:

$$E_b = E_0 \exp(-kx) + E_\infty \qquad (1)$$

where x represents the specimen thickness and E_0 and E_∞ represents the limiting breakdown strength as $x \Rightarrow 0$ and ∞ respectively. Here, E_0 = 172 ± 6 kV mm^{-1} and E_∞ = 32 ± 5 kV mm^{-1}. Since the Weibull distribution is essentially a weak link theory, it is tempting to ascribe the breakdown value in the limit

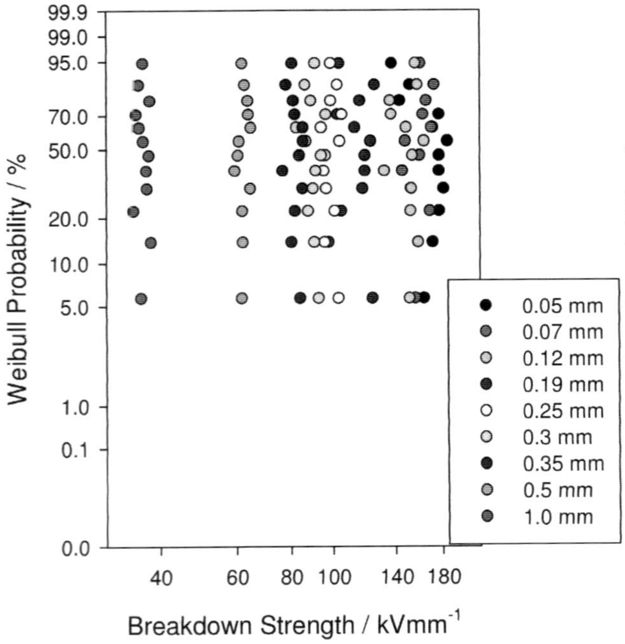

Figure 8. Weibull plots of breakdown data for epoxy samples of different thicknesses

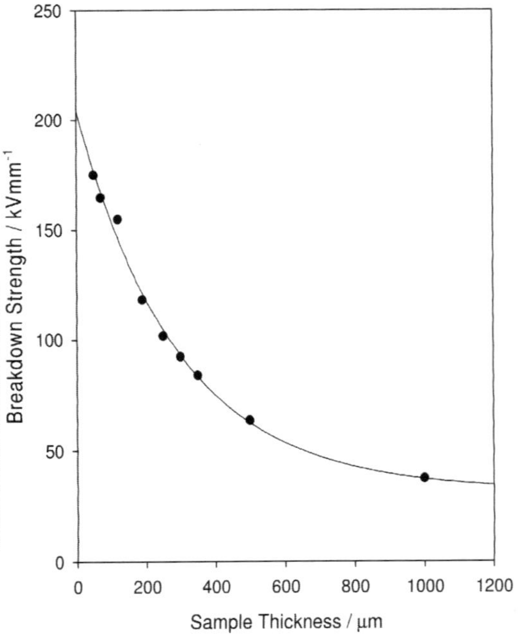

Figure 9. Variation of breakdown strength with thickness of the unfilled epoxy

that $x \Rightarrow 0$ as corresponding to a defect-free specimen, such that E_o will represent some form of intrinsic breakdown strength; the value of $E_0 = 172 \pm 6$ kV mm^{-1} derived from figure 9 is, however, significantly lower than "intrinsic breakdown strength" values in the literature [11].

5. Conclusions

It is evident that the mixing protocol used to prepare epoxy-based specimens can have a major effect on breakdown performance. In the case of nanodielectrics, this has traditionally been ascribed to nanofiller dispersion effects. The fact that comparable effects are seen here in unfilled epoxy systems is unexpected; evidently, good mixing of the resin and hardener is also critical and, it would appear that this can also benefit from sonication. Regimes A and B result in materials which exhibit reduced Weibull parameter values in all cases; sonication results in materials which exhibit enhanced Weibull parameter values in all cases. The thickness dependence of the breakdown strength conforms well to a simple exponential function, which enables the zero-thickness breakdown strength to be evaluated. The conventional interpretation given for the decrease in experimental breakdown strength with sample thickness is related to the probability that the tested volume contains an appropriate defect. The relatively low value of E_0 determined here suggests that, while the Weibull distribution may be an adequate description of breakdown data, its mechanistic validity is questionable.

6. References

[1] Rothon R N 2002 *Smithers Rapra Publisher,* volume 12, no. 9
[2] Huang X, Lewis S and Brittain W J 2000 *Macromolecule,* **33** 2000
[3] Tee D I, Mariatti M, Azizan A, See C H and Chong K F 2007 *Compos. Sci. Tech.* **67** 2584
[4] Chiang C L, Ma C C M, Wang F Y and Kuan H C 2003 *Euro. Poly. J.* **39** 825
[5] Jawahar P and Balasubramanian M 2006 *J. Nanomaterials* **1** 4
[6] Dissado L A and Fothergill JC 1992 *Electrical Degradation and Breakdown in Polymers*, ed G C Stevens (London: Peter Peregrinus)
[7] Reading M, Xu Z, Vaughan A S and Lewin P L 2011 *EIC Conf. Proc.* to be published
[8] Singhal R G *Thesis at Graduate Faculty of North Carolina State University*
[9] Reading M *Thesis at University of Southampton*
[10] Vaughan A S, Swingler S G and Zhang Y 2006 *Trans. Instit. Elec. Engin. Jpn* **126** 1057
[11] Herzig R and Baker W E 1993 *J. Mater. Sci.* **28** 6531
[12] Heylen A E D and Postoyalko V 2003 *IEEE Trans. Diel. Electr. Insul.* **10** 708

Laser ablation of silicone composites

I L Hosier[1]*, M S Abd Rahman[1], A S Vaughan[1], A Krivda[2], X Kornmann[2], L E Schmidt[2]

[1]ECS, University of Southampton, Highfield, Southampton, SO17 1BJ, UK

[2]ABB Switzerland Ltd, Corporate Research, 5405 Baden-Daettwil, Switzerland

*E-mail: ILH@ecs.soton.ac.uk

Abstract. Silicone rubber based composites are widely used to produce outdoor insulators. In adverse weather conditions these can be damaged by surface discharge activity resulting in deterioration of their dielectric properties and an eventual need for replacement. Inclined plane tracking tests are frequently used to determine the relative performance of different material formulations used in their construction. An alternative approach to characterisation is to use an infra-red laser to deposit known amounts of energy at a known rate to the sample's surface allowing comparative ranking of different materials. In this paper several silicone based composites have been ranked using a laser ablation technique and the results were then compared to those obtained from tracking tests on the same materials. The comparison indicates that laser ablation ranks the materials in the same order as the tracking tests and may therefore constitute a quick and cost effective method for the routine characterisation of outdoor insulation components.

Introduction

Silicone rubber based composites are used widely for outdoor high voltage components such as bushings, cable strings and as weather resistant coatings. In adverse weather conditions, dry band discharges [1] can occur resulting in tracking and surface damage which can ultimately result in dielectric or mechanical failure. Traditionally these components are filled with either aluminium hydroxide (ATH), silica or a mixture of the two, to improve their tracking resistance and erosion performance and to reduce the amount of expensive silicone material [2-4]. It is important in designing such materials that good arc resistant composites are chosen in order to maximize lifetime and minimize downtime and maintenance costs. Inclined plane tracking tests [2] have proved useful in determining the relative resistance of different material formulations to surface damage and have indicated that materials having a high filler loading are beneficial up to the point where a good dispersion of the filler can be maintained [3]. An alternative test method is to use a high power laser to apply a known amount of energy at a known power to a localized region of the sample [4-6]. Significantly, it has been demonstrated that laser ablation can give the same ranking as inclined plane tracking tests [4] and is considered superior to traditional FTIR analysis [5].

In the current investigation we have considered the role of the filler content and composition on the ablation resistance of several silicone rubber based composites. A range of laser powers and energies was employed and the resulting surface damage was quantified by pit-depth and mass loss measurements supplemented by optical microscopy. The various samples were then ranked and the results were compared to those obtained from inclined plane tracking tests on the same materials.

1. Experimental

1.1. Materials
Table 1 provides a list of the six samples used in this investigation arranged in order of increasing silica content and overall filler content. All the samples, with the exception of L, employ a conventional high temperature vulcanized (HTV) silicone rubber as the polymeric matrix. Sample M employs ATH as filler, sample H is its silica filled analogue whereas sample I contains a mixture of the two fillers. Samples F and E are silica filled and both contain commercial fire retardants to ensure compliance of these materials with UL94 V-0 classification [3]. Sample L is described as an "unfilled liquid silicone rubber".

Table 1: Samples used in this investigation, filler content is in parts per hundred rubber (phr)

Designation	Filler content
Sample L	None
Sample M	100 phr aluminium hydroxide (ATH)
Sample I	75 phr aluminium hydroxide (ATH) and 25 phr silica
Sample H	100 phr silica
Sample F	100 phr silica and 15 phr melamine cyanurate (MCA)
Sample E	100 phr silica and 30 phr ammonium polyphosphate (APP)

1.2. Sample characterisation
A carbon dioxide laser (Synrad) operating at a wavelength of 10.5 – 10.6 μm was used to irradiate the samples [6]. The power output of the laser was controlled by a PWM controller over the range 2 to 28 W. For safety, the laser was completely enclosed in a metal interlocked cabinet and the system was used under a fume extractor vented externally. At the sample (~20 cm away from the laser aperture), which was mounted horizontally to minimize any flow effects, the laser spot diameter was 3 mm. Each sample was weighed on a digital balance (accuracy 0.1 mg) before and after irradiation, to determine the mass loss after knocking out any loose filler. The pit depths were measured on bisected holes using a magnified hand graticule (accuracy 0.1 mm). Thermogravimetric analysis (TGA) was performed on ~ 30 mg samples using a Mettler Toledo TGA/DSC 1 scanning at 10 K/min under either dry nitrogen or an "air mixture" composed of 40 % oxygen and 60 % nitrogen.

2. Results

2.1. Thermogravimetric analysis(TGA)
The TGA results are shown in Figure 1. Sample L undergoes a gradual decomposition in nitrogen but rapidly combusts in the air mixture upon reaching 500 °C. Samples M and I show a behavior which is consistent with the decomposition of ATH (~380 °C), followed by the polymeric matrix [6] under both atmospheres. Whilst sample H heated under nitrogen indicates a slow decomposition of the polymeric matrix, the same sample under the air mixture undergoes rapid combustion at ~500 °C. The results confirm that ATH is an effective combustion retardant whilst silica is not [3]. The final two samples F and E behave similarly in both atmospheres confirming that both MCA and APP are performing as effective combustion retardants. For sample E the traces are consistent with decomposition of the polymeric matrix followed by the APP which decomposes into ammonia (NH_3), water (H_2O) and polyphosphoric acid [7]. In sample F the MCA appears to decompose first, into gaseous products such as nitrogen and ammonia [8] followed by the polymer. The mass losses due to the decomposition of the APP and MCA (12 % and 8 % respectively) are in line with the volume concentrations indicated in Table 1 and indicate that any polyphosphoric acid is eventually lost from sample E. In the analogous reaction ATH ($Al(OH)_3$) is converted to alumina (Al_2O_3) and water resulting in an overall mass loss of 34 %. The corresponding mass loss during the ATH decomposition in samples M and I respectively is

15 and 12 % indicating an initial ATH content of 47 and 35 % which is again consistent with the data shown in Table 1.

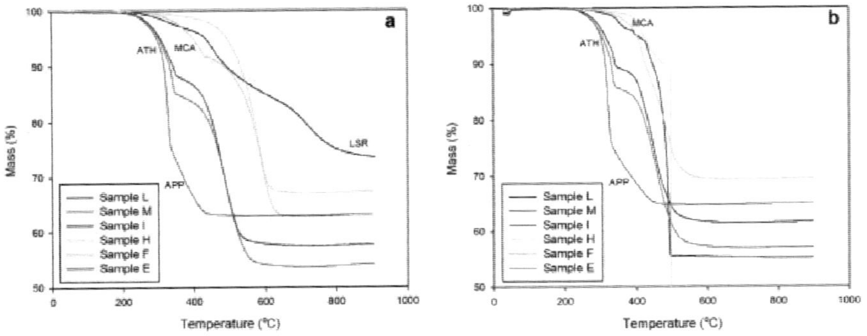

Figure 1: TGA mass loss traces for samples in (a) nitrogen (b) synthetic air.

2.2. Laser ablation results

Typical examples of the "as collected" data are shown in Figures 2 and 3 along with fitted exponential curves. Sample L, as an unfilled material, was found to offer the least resistance to ablation giving a maximum pit depth of 3.2 mm and a relatively high mass loss (Figure 2). By comparison, the filled materials (Figure 3) offered a greater resistance to ablation and gave lower pit depths and reduced mass loss. For all samples, the pit depth and mass loss are non-linear functions of energy, this primarily comes about through the formation of an ashy layer (Figure 6b) which serves to protect the underlying material from damage. At laser powers of 15 and 28 W the ablation behavior for any given energy is identical and more importantly, is independent of laser power, whereas at 3 W the pit depth and mass loss are lessened, presumably through the effects of lateral heat losses [6].

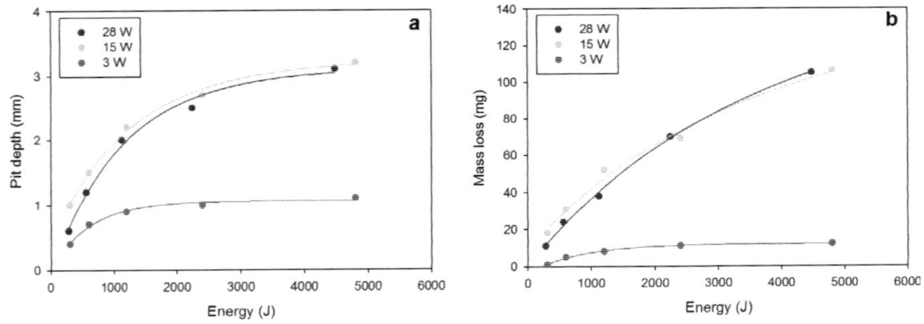

Figure 2: Ablation data for sample L (a) pit depth, (b) mass loss

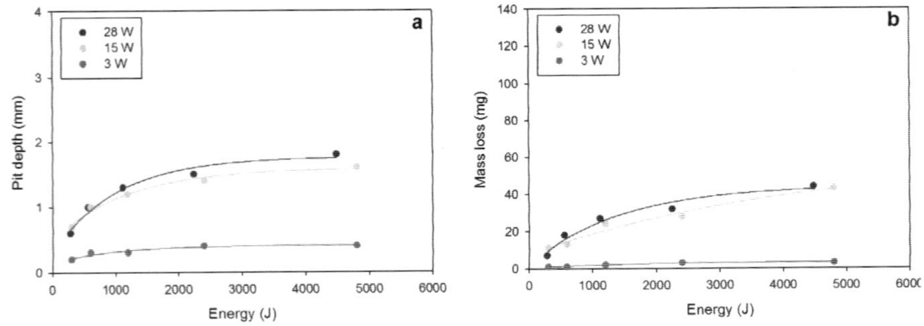

Figure 3: Ablation data for sample F (a) pit depth, (b) mass loss

2.3. Ranking

To rank the various materials both the pit depth and mass loss were compared at fixed laser power. Whilst the results at 3 W indicated smaller differences between samples (Figure 4) than higher laser powers (Figure 5) all the results indicated a common ranking scheme, from the worst to the best – L E

M I F H. These broadly follow the order established in Table 1 with one dramatic exception, sample E - despite the high filler loading of this sample and the addition of APP, this sample appears to behave much worse than expected. Considering samples M, I and H, it is clear that silica provides a composite with a better degree of ablation resistance than ATH. Finally sample F appears to perform somewhat worse than sample H in ablation tests despite the presence of the MCA fire retardant. The current results would indicate that APP, and to a lesser extent MCA, appear to degrade the ablation performance of the silica based composites.

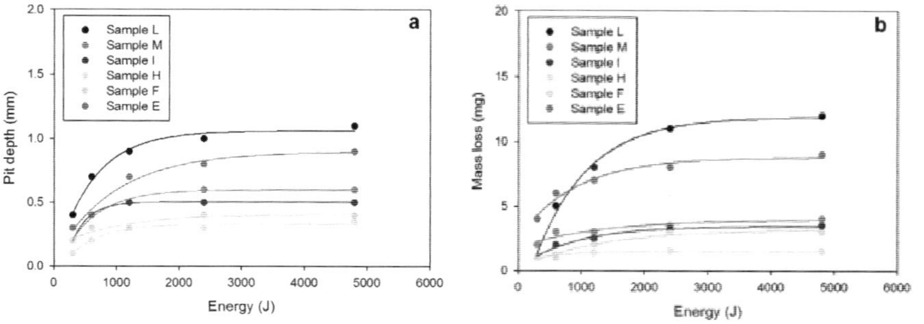

Figure 4: Comparison of materials at 3 W power (a) pit depth, (b) mass loss

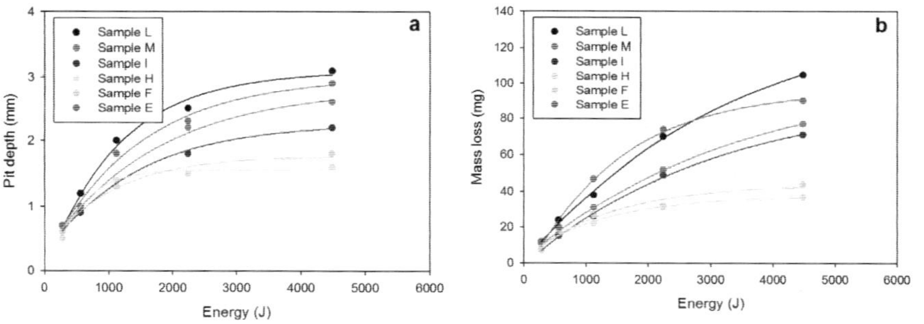

Figure 5: Comparison of materials at 28 W power (a) pit depth, (b) mass loss

2.4. *Morphology*

Figure 6 shows the effect of laser power on pit profile, at low laser power (Figure 6a) a small, shallow pit (~0.5 mm deep) is formed whereas at high laser power and the same energy (Figure 6b) a much deeper (~2.2 mm) and wider pit is formed. In the majority of the samples the pits are hemispherical in shape and are ashy which are the characteristics previously associated with filled composites as opposed to the "V" shaped profiles which are characteristic of unfilled materials [6].

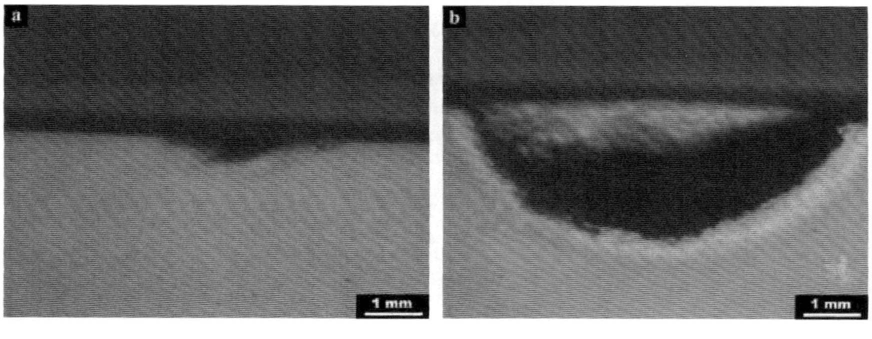

Figure 6: Comparison of pit profiles at different laser powers in Sample I (a) 3 W, 1600 s (b) 28 W, 160 s

Figure 7 shows a series of images of bisected pits formed at 15 W for 320 s comparing the effects of the different materials. Sample L shows broad, deep pits with little evidence of filler formation after irradiation, as expected (Figure 6a); the pit is partially filled with "broken" slabs of material giving an irregular hole profile. The remaining samples, by comparison, show a hemispherical profile with evidence of ash inside the ablated pits. Sample E assumes broad, deep pits (Figure 6b), whereas samples M, I, F and H in turn assume increasingly narrow and shallower pits. These samples contain a high filler loading (100 phr or above) which doubtless contributes to the ashing seen after ablation. It is not clear from the photographs why sample E should behave in such an unexpected way which suggests that the reason is chemical rather than physical in origin. The TGA data of sample E indicates that the polymeric matrix is degrading at a much lower temperature than any of the other composites considered; i.e. that the APP may be catalyzing its degradation in some way.

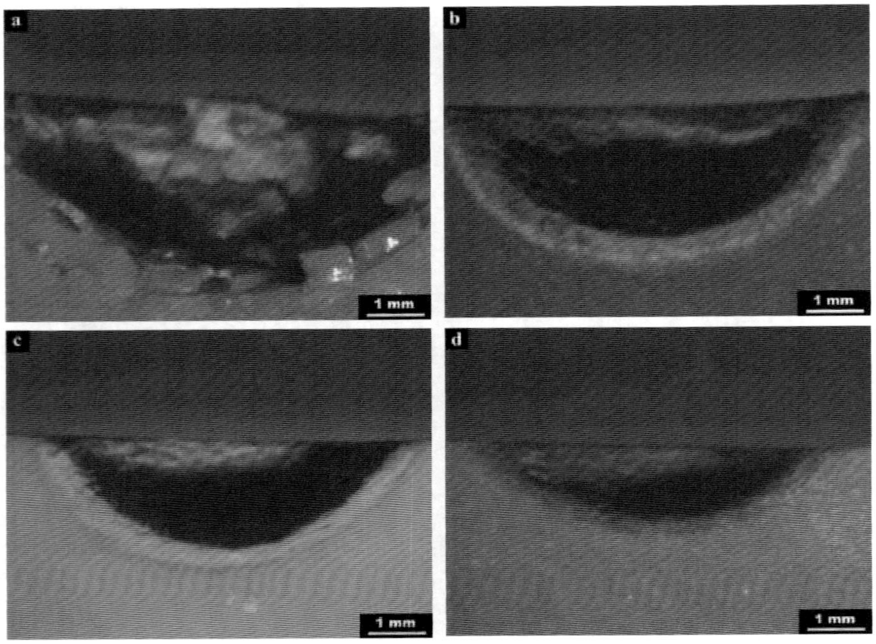

Figure 7: Images of pits formed at 15 W for 320 s, (a) Sample L, (b) Sample E, (c) Sample M, (d) Sample F.

2.5. Comparison to tracking tests

Table 2 shows the tracking test and breakdown results [3] arranged in order of decreasing track depth (i.e. improved performance). The tracking depth measurements do appear to correlate to the general ranking scheme established from ablation testing [4] with the exception of samples H and F which are reversed. However, both the tracking and ablation measurements from these samples are very close so experimental uncertainties could be the origin of this minor discrepancy. The correlation between the two sets of data suggests that laser ablation is a reliable way to rank different materials and may provide a quicker and more cost effective alternative to tracking tests.

Table 2: Tracking data, arranged in order of decreasing track depth

Designation	UL94 class	Failed at 4.5 kV	Average track depth
Sample L	Failed	0/20 (0 %)	3.8 mm
Sample E	V0	5/5 (100 %)	3.5 mm
Sample M	V0	3/15 (20 %)*	2.6 mm
Sample I	V0	5/15 (33 %)	2.1 mm
Sample H	Failed	5/15 (33 %)	1.3 mm
Sample F	V0	0/15 (0 %)	0.9 mm

*all samples passed 3.5 kV test with track depths of <0.5 mm

Significantly, composites utilizing silica in place of ATH perform much better in ablation and tracking tests but at the cost of reduced breakdown performance and loss of UL94 compliance. Silica based composites therefore require the addition of a suitable fire retardant to meet UL94 requirements. The presence of APP (Sample E) appears to be particularly deleterious in terms of all three test methods and should therefore be avoided; the TGA data suggests that it may be promoting the degradation of the polymeric matrix material. Finally, it is of no surprise, given the TGA data (Figure 1b), that both samples L and H fail the UL94 tests.

3. Conclusions

Comparisons between tracking and ablation tests reveal a good correlation, with all the filled samples behaving better than an unfilled LSR (liquid silicone rubber). In one case, where discrepancies do occur, the ablation and tracking test results were very similar indicating that experimental uncertainties may be a factor.

Composites employing silica, as opposed to ATH, appear to provide better resistance to tracking and ablation but at the expense of reduced breakdown performance and loss of UL94 compliance. Fire retardants are added to overcome this disadvantage but appear to reduce the tracking and ablation resistance, particularly APP (ammonium polyphosphate) which appears to promote the degradation of the polymeric matrix. A composite containing 100 phr silica in isolation provided the best resistance to tracking and with the addition of MCA (melamine cyanurate) to ensure UL94 compliance, would provide a useful composite material despite the MCA leading to slightly reduced tracking and ablation resistance.

4. Acknowledgements

We would like to thank ABB for providing the samples, tracking and breakdown data and for giving permission to publish this work.

5. References

[1] Srinivasan D, Karady G G, 2004 *IEEE Trans. Pow. Del.*, **19**, 1746-50.

[2] Krivda A, Schmidt L E, Kornmann X, Ghorbani H, Ghorbandaeipour A, Eriksson M, 2009 *IEEE Electr. Insul. Mag.*, **25**, 14-22

[3] Schmidt L E, Kornmann X, Krivda A, Hillborg H 2010, *IEEE Trans. Diel. Electr. Insul.*, **17**, 533-40.

[4] Meyer L H, Jayaram S H, Cherney E A 2005, *IEEE Trans. Diel. Electr. Insul.*, **12**, 1201-1208.

[5] Pinnangudi B, Gorur R S, Poweleit C D 2006, *IEEE Trans. Diel. Electr. Insul.*, **13**, 423-429.

[6] Hosier I L, Abd Rahman M S, Westenbrink E W and Vaughan A S, 2010, *Proc. 2010 IEEE International Conference on Solid Dielectrics*, 455-59.

[7] Song L, Wu K, Wang Y, Wang Z, Hu Y, 2009, *J. Macrom. Sci.: Part A: Pure Appl. Chem.*, **46**, 290-95.

[8] Liu Y, Wang Q, 2009 *J. Polym. Res.*, **16**, 583-589.

Numerical modelling of needle-grid electrodes for negative surface corona charging system

Y Zhuang, G Chen and M Rotaru

School of Electronics and Computer Science, University of Southampton, UK

E-mail: yz205@ecs.soton.ac.uk

Abstract. Surface potential decay measurement is a simple and low cost tool to examine electrical properties of insulation materials. During the corona charging stage, a needle-grid electrodes system is often used to achieve uniform charge distribution on the surface of the sample. In this paper, a model using COMSOL Multiphysics has been developed to simulate the gas discharge. A well-known hydrodynamic drift-diffusion model was used. The model consists of a set of continuity equations accounting for the movement, generation and loss of charge carriers (electrons, positive and negative ions) coupled with Poisson's equation to take into account the effect of space and surface charges on the electric field. Four models with the grid electrode in different positions and several mesh sizes are compared with a model that only has the needle electrode. The results for impulse current and surface charge density on the sample clearly show the effect of the extra grid electrode with various positions.

1. Introduction

Over the years, considerable interest has been shown in the surface potential decay of corona charged polymeric materials. It is a simple and low cost tool used to examine electrical properties of dielectrics. The study of the potential decay in dielectric materials has a long history and is closely related to the wide application of corona charged dielectrics. It became a very popular topic after Ieda et al found the special cross-over phenomenon during surface potential decay in 1967 [1]. As a result many assumptions and hypothesis have been made to explain the phenomenon. There was also significant improvement in experimental equipment and environment. Baum et al [2] firstly introduced the needle-grid corona charging system shown in figure 1 and compared the surface charge density on the dielectric material with needle only corona charging system experimentally. Recently most of the corona charging systems are using the needle-grid set-up mainly for two reasons: firstly the surface potential of the dielectrics can be easily controlled by adjusting the magnitude of the grid electrode potential, and secondly the capability of this set-up to get a uniformly distributed charge density along the dielectric surface [3]. However, there is no work and available data on the potential differences and distance between the two electrodes such that a perfectly charged sample would be achieved. It has been found that there is no consistent data available to set-up up the grid position in relationship with both needle and sample. For example, in the literature available, the grid position is varied from 0.5 cm to 1.5 cm above the dielectric samples and the needle position is moved between 1 cm to 5 cm above the dielectrics [4, 5, 6].

The model in this work is based on a gas discharge model [7]. It was built in 2D axial symmetry configurations using the finite element COMSOL Multiphysics package. It has been proved that the negative corona discharge occurs in the vicinity of the cathode, therefore, due to the high electric field and charge densities, a very fine 'FEA mesh' need to be used around the needle head and along the axial symmetry line (figure 2). In this paper, the effect of adding the grid electrode to the corona charging system will be discussed by studying the impulse current formed and dielectrics surface charge density on sample surface.

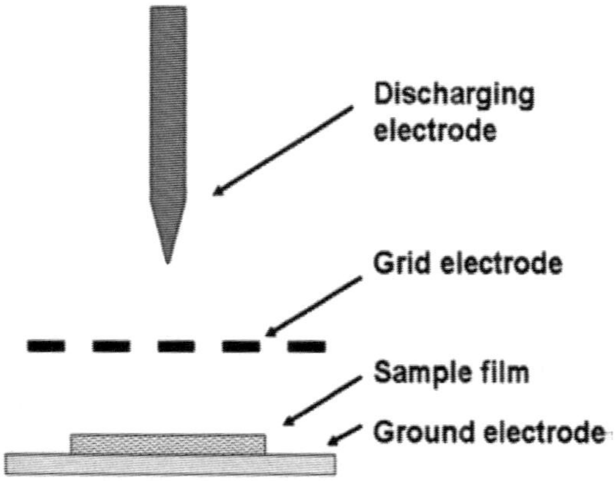

Figure 1. Needle-Grid Electrodes Corona Charging System [8].

Figure 2. 'FEA mesh' distribution in the model.

2. Model Description

2.1. Geometry of the model

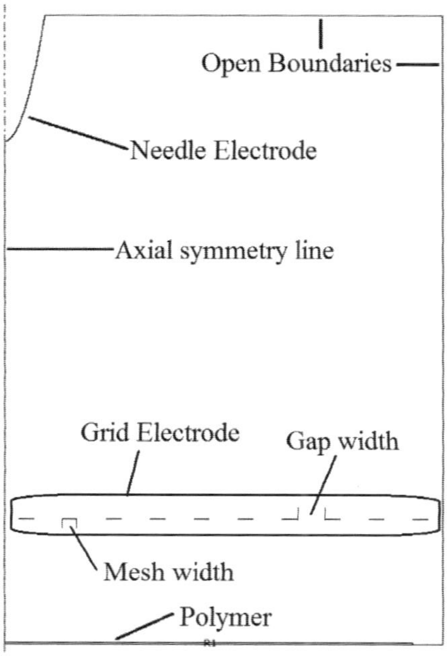

Figure 3. Corona Charging system simulation geometry.

From figure 3, it can be observed that the needle in our models has a curvature. The geometry of the needle is drawn with the help of equation (1). The radius of the curvature at the needle tip is 174 μm. In the model shown on figure 2 the sample has a diameter of 28 mm and it is considered to be a polyethylene film with 50 μm thickness. In the first instance, the grid electrode is modelled as several concentric circular conductors, which are varied at different radiuses from the axial symmetry line of the model. The grid electrode has a thickness of about 500 μm; however in our model, this dimension is reduced to zero to limit the size of the 'FEA mesh', therefore reducing the computer memory. The bottom surface of the polyethylene sample is grounded and the needle electrode is fixed at 2 cm away from the ground. The grid electrode is set 0.5 cm or 1 cm away from the ground respectively. The model is solved using a transient solver within COMSOL. The electric potential of the needle and grid electrode are initially set to 0 V. Within the first 0.1 ns of the simulation these potentials are brought to the working voltage levels -6500 V for the needle and -1000 V for the grid respectively. From 0.1 ns to 5 μs when the simulation is stopped, the two voltage levels are kept constant

$$\left(\frac{z}{b}\right)^2 - \left(\frac{r}{a}\right)^2 = 1 \; ; \; \frac{a^2}{b} = radius\ of\ curvature \tag{1}$$

2.2. Mathematical model

A well-known hydrodynamic drift-diffusion model consists of a set of continuity equations coupled with Poisson's equation, given as:

$$\frac{\partial N_e}{\partial t} = N_e\alpha|W_e| - N_e\eta|W_e| - N_eN_p\beta_{ep} - \nabla\cdot(N_eW_e - D_e\nabla N_e) \tag{2}$$

$$\frac{\partial N_p}{\partial t} = N_e\alpha|W_e| - N_eN_p\beta_{ep} - N_nN_p\beta_{np} - \nabla\cdot(N_pW_p) \tag{3}$$

$$\frac{\partial N_e}{\partial t} = N_e\eta|W_e| - N_nN_p\beta_{np} - \nabla.(N_nW_n) \tag{4}$$

$$\nabla\cdot(\varepsilon_r\nabla V) + \frac{e}{\varepsilon_0}(N_p - N_e - N_n) = 0 \tag{5}$$

where t is time, e the electronic charge, ε_o the dielectric constant of free space, ε_r the relative permittivity and V the electric potential; subscripts e, p, n represent electrons, positive and negative ions respectively; N_e, N_p and N_n are the ion number density; W_e, W_p and W_n the electron, positive and negative ion drift velocities; α, η, β and D the ionization, attachment, recombination and electron diffusion coefficients respectively. The simulation parameters in equations (2-5), which are functions of the local electric field, are given in table 1.

Table 1. Simulation Parameters [7].

Parameters	Functions (Electric field strength $E = \varepsilon_r\varepsilon_0\nabla V$ in $V\cdot cm^{-1}$)
$\alpha(cm^{-1})$	$3500\exp\left(-1.65\times10^5 E^{-1}\right)$
$\eta(cm^{-1})$	$15\exp(-2.5\times10^4 E^{-1})$
$\beta_{ep} = \beta_{np}(cm^3 s^{-1})$	2×10^{-7}
$W_e(cm\ s^{-1})$	$-6060E^{0.75}$
$W_p(cm\ s^{-1})$	$2.7E$
$W_n(cm\ s^{-1})$	$-2.7E$
$D\left(cm^2 s^{-1}\right)$	1800

The total current can be then computed using the energy conservation law:

$$I = \frac{1}{V_{needle}}\left(\frac{d}{dt}\int_V W dV + \int_V P_D dV\right) \quad (6)$$

where $W = \frac{1}{2}\varepsilon_0\varepsilon_r E^2$ and $P_D = \mathbf{E}\times\sum eN_i\mathbf{W}_i$.

The surface charge density can be calculated by integrating the normal component of charged particle current densities at the surface. In cylindrical coordinates, this can be expressed as

$$\sigma_s(t) = \int_0^t (J_{ez} + J_{nz} - J_{pz})dt \quad (7)$$

where $J_{iz} = eW_{iz}N_i$.

2.3. Boundary and initial conditions

The boundary conditions of all the models are shown in table 2.

Table 2. Boundary Conditions for all the models.

Application mode	Convection and diffusion N_e	Convection and diffusion N_p	Convection and diffusion N_n	Electrostatics V
Axial symmetry line	Axial symmetry	Axial symmetry	Axial symmetry	Axial symmetry
Needle electrode	Flux	Convective flux	Concentration=0	V = -6500
Grid electrode	Convective flux	Convective flux	Concentration=0	V = -1000
Polymer upper surface	Convective flux	Insulation/Symmetry	Convective flux	Surface charge
Open boundaries	Convective flux	Convective flux	Convective flux	Zero charge/Symmetry
Outer ground boundary	Convective flux	Concentration=0	Convective flux	$V=0$

At the needle electrode, the boundary condition for electrons is given as a secondary emission flux when positive ions strike the cathode

$$N_e|\mathbf{W}_e| = \gamma N_p|\mathbf{W}_p| \quad (8)$$

where γ has been used in the range from 10^{-2} to 10^{-3}, it is chosen as 0.01 in this model [7].

The initial conditions are given as

$$N_e = N_{\max}\times\exp\left(-\frac{(r-r_0)^2}{2s_0^{\ 2}} - \frac{(z-z_0)^2}{2s_0^{\ 2}}\right) \quad (9)$$

where $N_{max} = 10^{15}$ m^{-3}, r_o = 0 µm, z_o = 2 cm, s_o = 25 µm; $N_p = N_n = 0$ [7].

3. Simulation Results

Five different models with or without the grid electrode were built as shown in table 3. Every model was run for 5 µs and the impulse current and surface charge density was observed. In this section a comparison of the obtained results is presented.

Table 3. Description of the Models.

Model Number	Description
Model I	Needle Only
Model II	Grid electrode 5mm above the ground, mesh width is 5mm and gap width is 10mm.
Model III	Grid electrode 10mm above the ground, mesh width is 5mm and gap width is 10mm.
Model IV	Grid electrode 10mm above the ground, mesh width is 10mm and gap width is 10mm.
Model V	Grid electrode 10mm above the ground, mesh width is 10mm and gap width is 5mm.

3.1. Effect of the grid electrode height

In this section, the effect of the extra grid electrode with two different positions is compared with the model that has a needle electrode only. The results for models I, II and III are shown in figures 4 and 5.

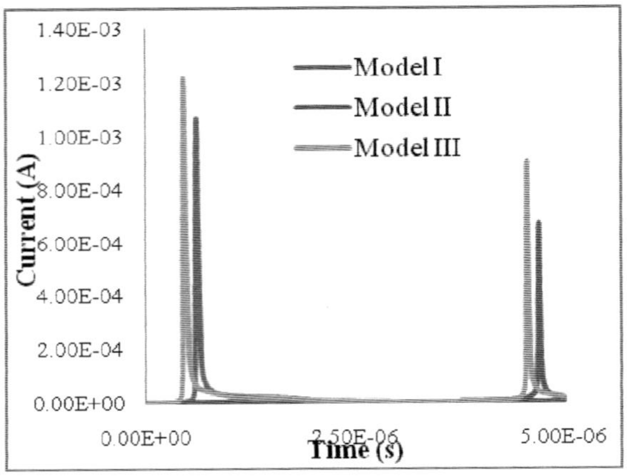

Figure 4. Current for models I, II and III.

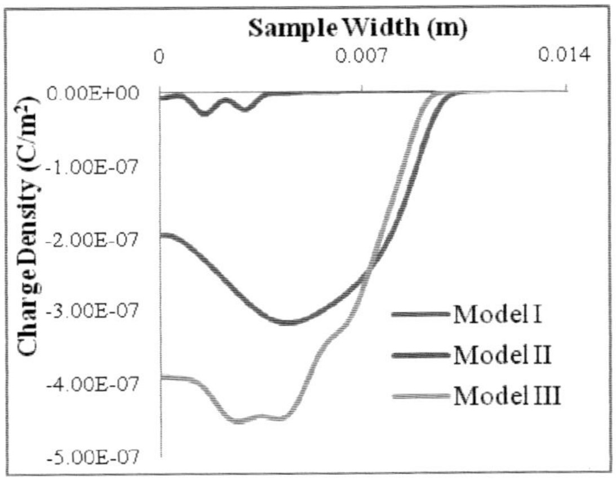

Figure 5. Surface charge densities for models I, II and III at 5 μs.

From the graphs above, it can be seen that the initial impulse current for model I and III shows at similar time, but it appears extremely late in model II. In contrast model III shows the existence of two impulse currents within 5 μs. Large differences in surface charge densities were observed. It can be noticed that at the point where the impulse current appeared, a great amount of electrons are generated as shown in figure 6. They will be pushed down to the sample surface, leading to an increase of surface charge density. As a result, model III has the largest surface charge density because of one more impulse current compared to the other two. It has been reported in [7] that the higher needle electrical potential will cause larger initial impulse current amplitude and a higher frequency of the Trichel pulse. In models II and III, the needle electrical potential is the same as in model I; however the impulse current is different. Therefore, the electric field underneath the cathode affects the behaviour of impulse current. From model I, it can be found that the electrical potential at 5 mm and 1 cm above the ground is about -600 V and -1200 V at the symmetry axis. The electrical potential at 5 mm above the ground is fixed to -1000 V in model II, therefore, reduces the electric field at the needle and leads to a smaller and slower first impulse current. In model III, the electrical potential is restricted to -1000 V at 1 cm above the ground, which leads to a higher electric field underneath the cathode and hence in this case a larger and faster first impulse current is observed.

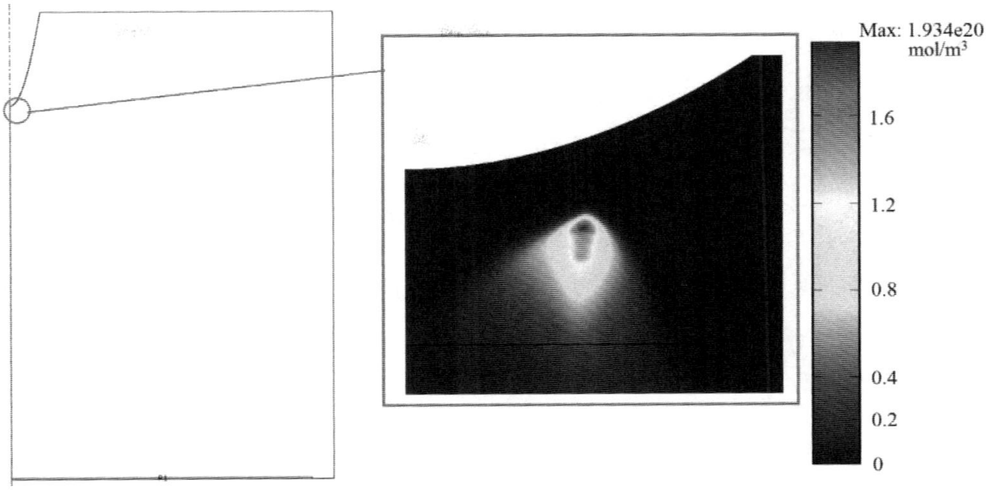

Figure 6. Electron density plot at 0.6 μs (peak of first impulse current in model I).

3.2. Effect of the geometry of grid electrode

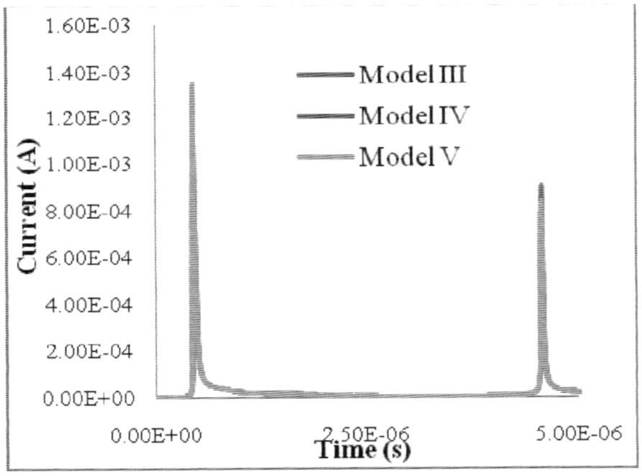

Figure 7. Current for models III, IV and V.

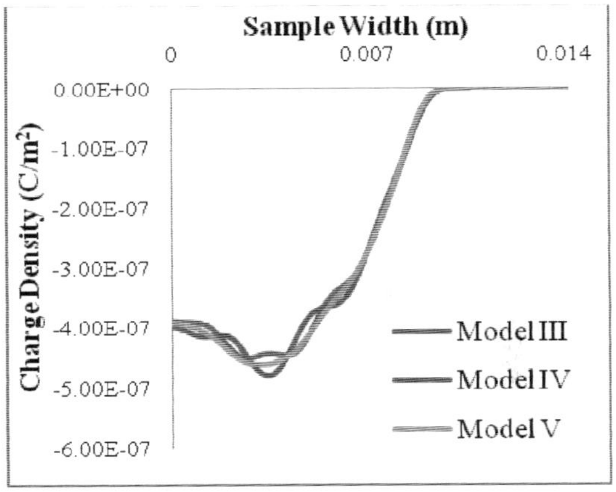

Figure 8. Surface charge densities for models III, IV and V at 5 μs.

The real grid electrode is a fine mesh which varies in width and gap, which is different from the one built in current model. Therefore, three different geometries (model III, IV and V in table 3) of the grid electrode were selected and modelled to observe any possible effects of the mesh geometry to the surface charge density at the polymer. The results for these models are shown in figures 7 and 8. From those graphs, it is clear to see that the difference between the surface charge density is tiny. From these models, it can be noticed that at the beginning of the corona charging process, the electrons will travel from the cathode to the sample as model I; however with the infinitely thin grid electrode in the current models, the electrons can pass through the mesh and mainly through the gaps of the mesh. In figure 9, the logarithm plot of the electrons is shown at three different simulation times: 0.01 μs, 0.5 μs and 1 μs. It is clear to see that at the beginning of the simulation, a cloud of electrons sits underneath the needle electrode and then they were pushed towards the polymer; the process of how electrons passing through the grid electrode can be also observed clearly from 0.5 μs and 1 μs results. Therefore, it can be said that a 2D-axis symmetry model can be used with confidence to simulate corona charging process and it is much more efficient than a larger 3D model.

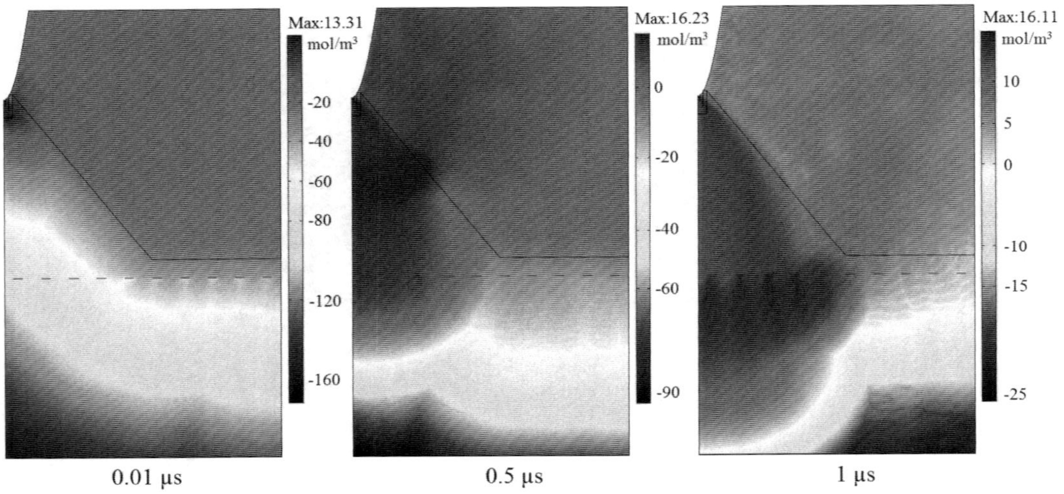

Figure 9. Logarithm plot of electrons for model III at different times.

4. Conclusion

It can be concluded that by adding the grid electrode into the corona charging system the whole system will be affected. Varying the distances between the grid electrode and the ground will result in a different impulse current and therefore make the surface charge density difference. Our simulation results showed that the different grid electrode geometries do not affect much the current or the surface charge density at 5 μs. However, much longer solving time might be necessary to observe any possible effects of the grid electrode, and our model should be useful for such future observations.

Acknowledgment

The first author would like to thank Dr. Trung Tran for his invaluable help on building the model.

References

[1] Ieda M, Sawa G and Shinohara I 1967 A decay process of surface electric charges across polyethylene film *J. Appl. Phys.* **6** 793–4

[2] Batra I P, Keiji Kanazawa K and Seki H 1970 Discharge characteristics on photoconducting insulators *J. Appl. Phys.* **41** 3416–22

[3] Xu Z, Zhang L and Chen G 2007 Decay of electric charge on corona charged polyethylene *J. Phys. D: Appl. Phys.* **40** 7085–9

[4] Ziari Z, Sahli S and Bellel A 2010 Surface potential decay of low density polyethylene (LDPE) films under different corona discharge conditions *M. J. CONDENSED MATTER*

[5] Zhukov S and Seggern H V 2010 Piezoelectrets from sandwiched porous polytetrafluoroethylene films with different porosity *2010 International Conference on Solid Dielectrics, July 4-9, 2010,* Potsdam, Germany

[6] Zha J, Chen G, Zhuang Y and Dang Z 2010 Novel surface potential decay of TiO_2-based polyimide nanocomposite films *2010 Annual Report Conference on Electrical Insulation and Dielectric Phenomena, 17-20 October 2010,* Purdue University, West Lafayette, Indiana, USA

[7] Tran T N, Golosnoy I O, Lewin P L and Georghiou G E 2009 Two dimensional studies of trichel pulse in air using the finite element method *2009 IEEE Conference on Electrical Insulation and Dielectric Phenomena, 18-21 october 2009,* Virginia Beach, Virgina, USA

[8] Chen G, Xu Z and Zhang L 2007 Measurement of the surface potential decay of corona-charged polymer films using the pulsed electroacoustic method *Measurement Science and Technology,* **18** 1453–1458

AUTHOR INDEX